DARWIN
ALL THAT MATTERS

DARWIN

Alison M Pearn

First published in Great Britain in 2015 by John Murray Learning. An Hachette UK company.

First published in US in 2015 by Quercus US

This edition published in 2015 by John Murray Learning

British Library Cataloguing in Publication Data: a catalogue record for this title is available from the British Library.

Paperback ISBN 978 1 473 60284 7

Ebook ISBN 978 1 473 60308 0

1

Typeset by Cenveo® Publisher Services.

Printed and bound in Great Britain by CPI Group (UK) Ltd., Croydon, CR0 4YY.

John Murray Learning policy is to use papers that are natural, renewable and recyclable products and made from wood grown in sustainable forests. The logging and manufacturing processes are expected to conform to the environmental regulations of the country of origin.

John Murray Learning
Carmelite House
50 Victoria Embankment
London
EC4Y 0DZ

www.hodder.co.uk

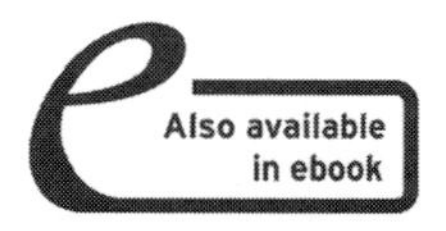

Contents

Acknowledgements

The author and publishers would like to thank the Syndics of Cambridge University Library for permission to reproduce the illustrations in this book. Quotations from Darwin's published correspondence are by permission of Cambridge University Press; quotations from Darwin's unpublished writings are by permission of William Darwin. The author would like to thank all her colleagues at the Darwin Correspondence Project, without whom this work would not be possible. The author is also grateful to the following friends for their invaluable comments on earlier drafts: Hugh Bennett, Shelley Innes, Marsha Groves, Anne Secord, Jim Secord, Sally Stafford and Charissa Varma.

1

Introducing Darwin

'... ignorance more frequently begets confidence than does knowledge: it is those who know little, and not those who know much, who so positively assert that this or that problem will never be solved by science.'

Charles Darwin, *The Descent of Man*, 1871

'The subjects discussed in this volume are so connected that it is not a little difficult to decide how they can be best arranged.'

Charles Darwin, *The Variation of Animals and Plants under Domestication,* 1868

Darwin is hard to escape. His name continually crops up in a range of cultural contexts from novels to films, and he is still constantly in the news. He is one of a handful of scientists so universally known that he has become more a brand than a person. His ideas were game-changers, fundamentally altering how we think of our own place in the natural world, but they are often misunderstood, misrepresented or misapplied. Two events – the *Beagle* voyage and the publication, more than 20 years later, of his book *On the Origin of Species* – are so famous that they tend to distort any balanced appreciation of his life and achievements.

The difficulty in explaining anything to do with Darwin is where to begin. Everything is indeed connected, which in a way is the whole point – Darwin is all about connections. He approached the practice of science as a seamless occupation, which perhaps explains how he was able to theorize so effectively about the common ancestry of all living things and arrive at the mechanism for species change that he called 'natural selection'. His investigations spanned the natural world – rocks, plants and animals. His research subjects included bees, dogs, corals, pigeons, worms, barnacles, orang-utans, babies, orchids, primulas, flax, corn, sundews and pitcher plants. He looked especially hard at the blurred margins: at plants that move and turn the tables on animals – the twiners, climbers, mimics and predators; at animals that behave like humans and humans that behave like (other) animals – altruistic gorillas and heartless people; at colonies of animals behaving like plants; at single organisms behaving like colonies; and at organisms with two sexual forms, or three, or none.

Darwin lived and worked surrounded by a close group of family and friends, and embedded in a wider network of scientific collaborators. His research became a social enterprise. His books were eagerly awaited, absorbed and discussed, and his readers in turn became co-workers, sending him observations on the natural world, solicited or not. Darwin as a 'lone genius' is one of the least-deserved reputations in science, undervaluing, as it does, both his own achievements and the contributions of others.

Darwin's early life

Charles Darwin's life spanned a period of remarkable change in politics, economics, technology and culture. He was born on 12 February 1809 into the Regency world of Jane Austen; her first novel was published when he was two. He was a near contemporary of Charles Dickens, who published his debut book in 1836, the same year Darwin returned from sailing round the world on a wooden ship, and just a year before a very young Princess Victoria ascended an unpopular throne. When Darwin died in 1882, aged 73, the imperialist adventure tales of Rudyard Kipling and H. Rider Haggard were about to appear, and a teenage H. G. Wells was on the verge of writing his first science fiction. The United Kingdom had been transformed into a politically stable, industrialized and prosperous country. It was powered by an iron web of steam railways, protected by an ironclad navy, and seated at the heart of a vast empire.

Childhood, 1809–25

Charles was the fifth of six children born to a prosperous doctor living in a large house just outside Shrewsbury, a market town on the border of England and Wales. His father, Robert Darwin, was the son of the famous polymath Erasmus Darwin, who had been a close friend of the chemist Humphry Davy and the poet Samuel Taylor Coleridge.

Erasmus Darwin (1731–1802)

Charles's grandfather was the author of a series of popular and influential books, including The Botanic Garden (1789–91), an account of recent scientific advances written in verse; *Zoonomia; or, the Laws of Organic Life* (1794–6), which was a medical book; and the posthumously published poem *The Temple of Nature* (1803). All had a strong evolutionary theme, imagining the gradual development of different forms of life over many generations, but without suggesting how this happened. The term 'Darwinism' was originally used – pejoratively – to refer to Erasmus's poetic style. Charles made a brief reference to his grandfather in the introduction to the third edition of *On the Origin of Species* (1861), and was later accused by the writer Samuel Butler of plagiarizing Erasmus's ideas.

Darwin's mother, Susannah, was a Wedgwood – the Wedgwood pottery company had been founded by her father and was run by her brother Josiah. As a child, Charles went with his father on house calls to patients who included the colonial hero Clive of India, made up secret codes with his younger sister, collected rocks and shells, and ganged up with his brother (also called Erasmus) against the three older girls.

Susannah Darwin died when Charles was eight years old and he was then brought up by his sisters, going away to school only as far as the local Shrewsbury Grammar. The curriculum was dominated by the study of Greek and Latin authors, with some Shakespeare, but nothing much that we would recognize as science. Charles doodled in his schoolbooks and ran home whenever he could. At 16, rather younger than most, he was sent to join his brother who was switching universities from Cambridge to Edinburgh. Both intended to follow their father and grandfather into medicine.

Edinburgh, 1825–7

There were no entrance qualifications and little formal structure at Edinburgh University. The intellectual climate was liberal. Charles joined the Plinian Society, which met to discuss natural science, and he became friendly with Robert Grant, a lecturer on zoology and an enthusiastic naturalist. Darwin's first academic study was of some tiny marine creatures collected during seaside walks with Grant. Grant was also an evolutionist, championing Lamarckism (named for the French naturalist Jean-Baptiste Lamarck), which denied divine design in nature and was regarded as dangerously close to atheism.

Darwin never got over his horror at the pain of surgery – anaesthetics had not yet been discovered – so after two years he abandoned medicine and left Edinburgh without completing his degree – not unusual at the time – and again followed his brother who had returned to Cambridge to complete his studies there.

Darwin and Unitarianism

Both the Darwin and Wedgwood families came from a Unitarian tradition, a Christian Protestant sect characterized by the promotion of religious tolerance and an emphasis on individual conscience, rather than dogma, as a guide to conduct. Unitarianism rejects the three-person nature of God and the divine nature of Jesus, both central to the '39 Articles' of the Church of England, which also assert the authority of the Bible and Church and the doctrines of original sin and justification by faith.

In England, Unitarianism was linked to political radicalism and outlawed until 1813. Before the repeal of the Test and Corporation Act in 1828, anyone wanting to hold a government post in England still had to accept the 39 Articles. Even then, Unitarians found it difficult to hold public office or pursue a profession well into Darwin's adult life, and although his mother and sisters remained Unitarians, he and his brother were both baptised into the Anglican Church.

Cambridge, 1828–31

Cambridge was a small and intimate microcosm of privileged society, and Charles was a typical student who did what students do: he partied, talked about girls, and got into debt. He was popular and sociable, organized hunting expeditions, formed a small dining club, and made lifelong friends. At the end of three years he was relieved to do reasonably well in the exams, but the interest his teachers took in him hints at more academic talent than his paper qualifications suggest.

Two Cambridge professors, in particular, helped promote the beginnings of a scientific career, the importance

▲ 'Go it Charlie!' As a student, Darwin was such a keen collector of beetles that his friend Albert Way drew these caricatures of him riding one. (Cambridge University Library MS DAR 204: 29)

of which no one at the time could have foreseen: John Stevens Henslow, professor of mineralogy and then of botany, and energetic director of the University Botanic Garden, and Adam Sedgwick, Woodwardian Professor of Geology and president of the Geological Society of London. Both men befriended Darwin. Henslow's lectures and field trips around Cambridge, not to mention his dinner parties, fostered Darwin's interest in natural history, and particularly in the systematic study of variation in plants. With Sedgwick, he had the chance to be part of some of the most significant geological investigations of the time, leading to the decoding of the oldest known fossil-bearing rock strata.

The Church as a possible career

'...a distant prospect of a very quiet parsonage...'

Letter to Caroline Darwin, 25–6 April 1832

It is sometimes said that Darwin was training at Cambridge to be a clergyman, or that he did a theology degree, with the implication that by not entering the Church he deliberately turned his back on a professed vocation. These claims are based on a misunderstanding both of the role of the Anglican clergy and of the university system in Victorian England. In the first half of the nineteenth century, half of all Cambridge and Oxford graduates went on to be ordained, but theology as a separate undergraduate degree subject did not exist at Cambridge before 1871. Until then, most

Cambridge students, including Darwin, got general or 'ordinary' degrees, which involved studying set texts in mathematics as well as in theology alongside a small number of classic literary works. There were really only three professions open to young men of Darwin's social standing: medicine, which he had already rejected, the law, and the Anglican ministry. Although he later briefly considered a university career, this never looked likely during his student days. There were precedents in his immediate family for choosing the Church – two of his cousins did so, including his contemporary at Cambridge and lifelong friend William Darwin Fox (a second cousin). Until the last year of the *Beagle* voyage Darwin was apparently resigned to the idea of becoming a clergyman, seeing it as a congenial way of life that would allow him to continue his pursuit of natural history.

The voyage of the *Beagle*, 1831–6

In August 1831 Darwin returned home from a geology field trip to Wales. His plans for a post-graduation expedition with friends to Tenerife had just fallen through, and he was at a loose end. Waiting for him was a letter from Henslow passing on an offer that came out of the blue. Robert FitzRoy, the captain of an Admiralty surveying vessel, HMS *Beagle*, was looking for a personable young man with an interest in natural history as a companion for his next voyage. As the Spanish empire in South America fragmented, Britain was determined to exploit the still largely unmapped continent, and the *Beagle*'s task was to chart the coast and explore the islands that lay along both the Atlantic and Pacific routes, looking for the natural resources and the safe harbours that would give Britain a commercial and military advantage.

The importance of this voyage in shaping Darwin as both a person and a scientist is so evident in retrospect that it is hard to remember that he was really just tagging along for the ride, and that neither its route nor its purpose in any way revolved around him. Right to the last months of the journey, a parsonage and a life of obscure dabbling in natural history still seemed, even to Darwin himself, the most likely outcome. He was not the first or even the second choice for the trip – Henslow had briefly considered going himself and then suggested his brother-in-law, Leonard Jenyns – and it was Darwin's social skills as much as his abilities as a scientist that got him the invitation.

Captains ran their ships like franchises: FitzRoy spent his own money on modifications to the *Beagle*, hired artists to record the voyage, and even for a time bought and paid for a second ship, the *Adventure*, to travel alongside the *Beagle*. Darwin was on board at FitzRoy's invitation and primarily to rescue FitzRoy, only four years older than himself, from social isolation. But he was also there on the clear understanding that he would use the opportunity to study the natural history of the places they visited and collect specimens of rocks, plants and animals, something usually done by the ship's surgeon. One of very few people in the whole of his life with whom Darwin did not get on, the *Beagle*'s surgeon left a few months into the voyage, his nose seriously out of joint.

The navy did not pay Darwin: the Admiralty fed him while on board ship (which, as he never got over seasickness, was as little time as he could get away with), but his father paid for everything else. Having supported his son through six years of on-again off-again university studies, Robert Darwin was understandably reluctant to continue to bankroll him for a further two (which in the event turned into five), and for a while the whole scheme seemed on the point of collapse. Only after some intense pleading from Charles did his father finally give his permission. After a couple of false starts, the *Beagle*, with Darwin on board, set sail on 27 December 1831.

'Whirled round the world ... in a ten-gun brig'

Letter to J. S. Henslow, 9 July 1836

It is hard to imagine now being on the open ocean in a wooden sailing ship measuring only 27.4 metres (90 ft) by 7.5 metres (24 ft 6 in), as one of 70-odd young men (and one young woman), without any form of communication other than letters sent through a haphazard network of other ships. They had few maps, and even fewer accurate ones. Darwin slept in a hammock slung across the chart table in a cabin about 3 metres (10 ft) square, shared with two other men. Over the next five years they encountered cholera, bandits, storms, riots and revolutions. The remnants of the indigenous population over much of South America was being wiped out in bloody fights with European settlers; even legitimate authority was often in the hands of men who were little better than warlords. The *Beagle*'s crew was called in to help quell violent unrest in Buenos Aires, came close to shipwreck in a storm off Cape Horn, and was almost stranded in the Antarctic when a calving glacier caused a tidal wave that threatened to sweep away their landing craft; Darwin was one of the first to realize the danger and ran to drag the boat further up the shore. They witnessed an erupting volcano and got caught in an earthquake; several of them died on the way, at least three from malaria. Darwin himself very nearly died of fever.

▲ **The *Beagle* and *Adventure* sketched by Conrad Martens off the coast of Chile, 2 July 1834 (Cambridge University Library MS Add 7984: 19)**

The ship's route took in Brazil, Argentina, Tierra del Fuego, Chile, the Falklands (twice), Tahiti, Australia, South Africa and the Cape – and an insignificant group of Pacific islands called the Galápagos. The *Beagle* followed a drunken path, going back over the same ground, driven back by storms or the need to make repairs, or simply to recheck survey measurements. They were nearly home when, to Darwin's howling despair, they doubled back across the Atlantic to resurvey their very first landfall in Brazil. Finally, on 2 October 1836, after a voyage of five years and two days, they landed back in England.

Darwin spent those five years observing, collecting and recording. He was often dropped at one point at the coast, rejoining the ship only after travelling many hundreds of miles on horseback. He had to organize these expeditions himself. He hired horses and guides, got rough maps drawn by settlers, hunted for his food, and cooked what he killed. He climbed the Andes from both sides, camped out, lassoed ostriches, eyed up the local women, and talked his way out of trouble when he lost his passport. He carried a notebook and a geological hammer, and kept detailed notes of the minerals, plants, animals, fish, birds and fossils he encountered. He devised numbering systems with different-coloured labels, and kept cross-referenced lists. Back on board, he copied out and expanded his notes and sent them back with letters to England, the nineteenth-century equivalent of a data backup – paper records were at high risk on a small ship in rough seas. He learned to preserve all sorts of specimens – from tiny marine worms, to cucumbers from the Galápagos, to steamer ducks from Tierra del Fuego – and sent them back in barrels and boxes to specialists in England. The ship's lieutenant complained about the clutter on his deck, already crowded with whaleboats and surveying equipment.

In Patagonia and in Tierra del Fuego, right at the southern tip of South America, the crew spent time with some of the most remote human groups in the world. The woman on board the *Beagle* was a Yaghan, a native of Tierra del Fuego, who, with two Yaghan men, was being returned home after several years living in England. FitzRoy had brought them back with him from a previous voyage, and hoped, wrongly as it turned out, that they would permanently imbibe Christian traditions and British cultural values.

Looking back, Darwin identified his experiences travelling the world with HMS *Beagle* as the most formative of his life. It gave him first-hand knowledge of the wide variety in terrain, plants, animals and people; it gave him an income of his own as a travel and science writer, and it helped him establish a network of scientific collaborators. He matured from a 22-year-old recent graduate whose most likely career path was in the Anglican Church to a young man recognized by the scientific establishment – but not for anything to do with evolution. There was no Eureka moment on the Galápagos, or elsewhere. It was for his contributions to geology rather than biology that he first became known. Although he had amassed specimens of many different species and was privately speculating on their development, his observations of large-scale rock formations, which were what first really excited him, and a brilliant new explanation for the formation of coral reefs were what made his name.

Charles Lyell (1787–1875)

Lyell was a Scottish geologist and the author of the book that probably shaped Darwin's thinking more than any other: *Principles of Geology*. Darwin was given the first volume, published the year before they sailed, by FitzRoy; he got the second in Montevideo in November 1832, and the third one was probably sent out to him by Henslow soon after it was published in 1833. Lyell championed the gradualist view that geological features are the result, not of ancient cataclysmic events – the biblical flood, for example – but of the gradual accumulation of small changes brought about by familiar everyday forces – weather, climate and seismic events. Darwin was an immediate and lifelong convert. Lyell was later equally impressed by Darwin's coral reef theory, and helped establish his scientific career. Although Darwin was frustrated by Lyell's later caution over evolutionary theory, the two men remained friends to the end of Lyell's life.

London, 1837–42

Darwin never left Britain again. He made just two more geological field trips, to Glen Roy in Scotland in 1838 and to North Wales in 1842; otherwise, everything he went on to do he did close to home. He established himself briefly in Cambridge, then in London, parcelling out his specimens for identification and widening his scientific contacts. One of the important consequences of his independence from the navy during the voyage was that his specimen collections remained his own property and he was the one who decided what research was done on them, and by whom. Just describing them took years: the fish went to Leonard Jenyns, the birds to the ornithologist and illustrator John Gould, and the fossils to the palaeontologist (and coiner of the word 'dinosaur') Richard Owen. Most of the plants were sent to the Royal Botanic Gardens at Kew, where they were eventually described by Joseph Dalton Hooker. Darwin kept just one collection for himself – the barnacles – but he did not start work on those for more than ten years.

Joseph Dalton Hooker (1817–1911)

A medical doctor and a botanist, Hooker travelled widely, building up plant collections, first as a naval surgeon and then on government grants. He met Darwin, already a hero of his, in 1844, and became his closest confidant and lifelong friend. More than a thousand of their letters still survive, and it was to Hooker that Darwin is first known to have revealed his theory of species change.

In 1855 Hooker became assistant to his father at the Royal Botanic Gardens, Kew, a government-run facility crucial to Britain's economic interests abroad, succeeding him as director in 1865. He not only provided Darwin with plants from all over the world but, as the centre of an efficient network of colonial botanists, diplomats and agricultural administrators, also introduced Darwin to many of the people who went on to help him with his work.

Darwin's initial inkling, towards the end of the *Beagle* voyage, that species might not be so stable after all, moved slowly to become a conviction. Over the next few years he recorded his ideas about how that change might take place in a series of small leather-bound notebooks marked 'Private'. From as early as 1837 he explored anything that might help answer the question 'What are species?' He studied the effects of selective breeding, hanging out in London pubs with pigeon fanciers, read widely, and systematically gathered data.

> *'As for a wife, that most interesting specimen in the whole series of vertebrate animals, Providence only know[s] whether I shall ever capture one or be able to feed her if caught.'*
>
> Letter to C. T. Whitley, 8 May 1838

He was secretary to the Geological Society for a while (the closest thing to an official post Darwin ever held), and he wrote and wrote and wrote, working first on a multi-volume report on the natural history of the *Beagle* voyage edited by FitzRoy, and publishing his own very successful travel diary. He briefly considered a university career, but decided instead on marriage, something Cambridge University fellows were not allowed to do until 1882.

He married a cousin, Emma Wedgwood, on 29 January 1839 and the couple moved into a rented house, 12 Upper Gower Street, in London. They lived on an allowance from Darwin's father, together with the income from Darwin's writing and rents on land, some his and some Emma's. Their first child, William, was born the following December, and Darwin immediately started making notes on this new specimen. Altogether the Darwins had ten children. Not untypically for the time, two died in infancy and another daughter, Annie, died aged ten in 1851.

Darwin on marriage

Around the time of his engagement, Darwin wrote out the pros and cons of marriage in two columns like a balance sheet. Among the benefits of getting married, he identified companionship in old age, the charms of female chitchat, and a 'nice soft wife on a sofa', which would be 'better than a dog anyhow'. As disadvantages, he listed the loss of time for reading, being forced to visit relatives, and missing out on going to America or up in a balloon. He put having children in both columns. His two versions of the list may have been intended to tease Emma, who kept them to the end of her life. They are still often read out at weddings.[1]

Down House and family

In 1842 the Darwins moved to Down House in the village of Down (now Downe) in Kent. Darwin described the three-storey house as 'oldish & ugly'[2] but with a good-sized study and plenty of bedrooms. The location, about 26 kilometres (16 miles) from central London, was carefully calculated to lie far enough outside the city to be quiet and cheap, but near enough for him to remain in touch with clubs and scientific societies. The household included a number of servants: a housekeeper, a butler, two footmen, gardeners, kitchen maids and a nursemaid. As the children got older they had governesses, including two German sisters whom Darwin co-opted to translate scientific papers for him.

The Darwins were one of the leading families of the district and became closely involved in parish life. Until she fell out with one of the vicars, George Ffinden (whom she privately referred to as 'the Ffiend'), Emma took the children to services at the Anglican parish church. There is no evidence that Darwin went to church, but the problems with Ffinden were more personal than theological, and both Darwins had been close friends with his predecessor, John Brodie Innes. The parish was the unit of local government, elementary education and philanthropy. Darwin served as a magistrate on and off from 1857 until 1874. Emma and the girls taught in Sunday school, and Charles was founder and treasurer of two parish charities. He kept some of the church accounts, contributed to the fund for replacing the organ, and worried about the damage done to the village church's reputation by the sharp financial practices and womanizing of a couple of the curates.

The Darwins were a close family and the children moved from being the subjects of research to lifelong collaborators. William, who went to work as a banker in Southampton, sent his father sketches of magnified pollen grains from the Isle of Wight; Henrietta observed orchids during her travels on the Continent, acted as unofficial secretary, and was a critic and editor of his books; George helped with maths and statistics; Francis was employed as his father's secretary and assistant for the last eight years of Darwin's life, and then became his first biographer. Darwin's wife Emma's role is less easy to establish simply because there is not much evidence – they were so rarely apart after their marriage that very few letters between them survive. She knew about his work, did not always agree with it, and worried about his lack of religious faith, but she helped him nearly every day with correspondence and was clearly far from passive.

The house was a place of lively debate and Charles's work was discussed among the family quite freely. As an 11-year-old, his son Horace understood Darwin's theories well enough to argue that if everyone killed adders the timid ones would run away and survive, and in time adders wouldn't sting any more – 'Natural selection of cowards!'[3] as his father summed it up. Henrietta at the age of 25 was described as 'quick and bright as steel',[4] a young woman of decided opinions, highly educated and well informed and perfectly happy to argue her point, not just with her father but with his scientific colleagues.

The family spent some time each year in London staying with relatives, and quite often had visitors themselves, but shortly after returning from South America Darwin began to suffer form a chronic illness that stayed with him for the rest of his life. As a result, he often cut visits short and rarely worked at writing or at experiments for more than a couple of hours a day. A visitor in 1868 described him as 'tall & thin, though broad framed', with a face that showed 'the marks of suffering & disease'; and yet he had 'the sweetest smile, the sweetest voice, the merriest laugh'.[5] He was particularly good at laughing at himself.

So what *was* wrong with Darwin?

Darwin's working life was notoriously shaped by his health, or lack of it. By the late 1840s he was widely known to be an invalid who could meet visitors for only perhaps a quarter of an hour before he had to withdraw. He had a succession of doctors, including the eminent Henry Bence Jones, an associate of Florence Nightingale. Emma kept a diary of his symptoms and in 1865 he sent a list to a popular medical practitioner, John Chapman, whose ice treatment Darwin tried. The list included wind, daily vomiting, spots before the eyes, dizziness, fatigue, hysterical crying, and nervousness whenever Emma left him.

Among the suggested causes of Darwin's symptoms are the blood disorder Chagas disease, which he could have contracted from an insect bite in South America; lactose intolerance, which a Victorian diet would certainly not have helped; an immune system damaged by exposure to the chemicals used to preserve specimens; or a mental disorder related either to the early death of his mother or to fear of the consequences of publishing his controversial theories. His own fear was that he had an inherited, and therefore heritable, condition, and he watched the health of his children nervously. Probably he was suffering from no one single condition, but health became an obsession for his entire family.

The road to *Origin*

Darwin did not publish his most famous book, *On the Origin of Species*, or *Origin* for short, until 1859, more than 20 years after his return from the *Beagle* voyage. By 1844 he was quite certain not only that species were not fixed, but that he had found the 'simple way' by which they became 'exquisitely adapted to various ends'. He told very few people. It was, he wrote to Joseph Hooker, a little tongue-in-cheek, 'like confessing a murder'.[6]

The parcel (bomb) under the stairs

After Darwin's death, a bundle of manuscripts was found in the cupboard under the stairs at Down House. The manuscripts turned out to contain Darwin's first attempts at articulating his theories. They show that by 1842 he already had a clear enough idea to write a 35-page outline – known as 'the pencil sketch' – and that two years later he had expanded this into a 230-page essay in which he used the term 'natural selection' for the first time. The 1842 sketch expressly excluded from its scope not just the 'first origin of life' but also of 'mind' and whether organic beings are descended from one ancestor or several. The 1844 essay is an almost complete outline of the argument as it was later published in *Origin*, with some of the most famous passages already articulated.

Marshalling great quantities of facts

Why didn't Darwin go public in the 1840s? The most important reason was strategic. He was determined to give his ideas the best possible chance of being taken seriously by both the scientific world and the world at large. As far as the public (or at least the establishment) went, the political climate was tense and any evolutionary ideas were all too easily associated with revolutionary ones, making them far less likely to be accepted. And in 1844, the year in which he wrote his essay, the less than favourable reception by his fellow scientists of an anonymous evolutionary work, *Vestiges of the Natural History of Creation*,[7] gave him serious pause.

Vestiges was widely read and caused quite a stir. The problem with the book was not its evolutionary thesis but its shaky scholarship, which seriously undermined its impact in scientific circles. That was a pitfall Darwin was determined to avoid. He was well aware that his ideas were controversial; he fully intended to publish in the end and face the consequences, but to be as well equipped to face them as possible. He set out to amass as much supporting data as he could, to think of as many counter-arguments as he could, and to work out how to demolish them. *His* theory was going to be stress-tested before it was released.

A painstaking taxonomic study of barnacles, which to outside appearances was his sole occupation for eight years from 1846 to 1854, was one way to establish his credentials as a solidly grounded natural scientist; you do not get much more solid than eight years of barnacle research. It paid off, earning Darwin the 1853 Royal Medal of the Royal Society. More importantly, his novel inclusion, in such a study, of fossil as well as living specimens allowed him to look at very long genealogies, and so to test his ideas about species and varieties. And it incidentally gave him vital insights into the origins of sex.

The 'Big Book'

By 1856 he was ready to start writing up his wider theories in earnest, and by 1858 he had reams of manuscript of a 'Big Book' that he intended to call *Natural Selection*. And it would have been a very big book indeed, one that 'very few would have had the patience to read'.[8]

And then it all blew up in his face.

Although he had begun cautiously to share the detail of his theory with only a very few close confidants after finishing the 1844 essay, it did become more generally known that he was working on 'the species question'. One person who knew the broad thrust of Darwin's research was Alfred Russel Wallace, a naturalist and explorer with whom Darwin started a correspondence in 1855 or 1856. This explains why it was to Darwin that Wallace sent a paper outlining his own evolutionary theory. It was June 1858, and Wallace was in the Malay Archipelago.

If Darwin's bundle of manuscripts under the stairs was an unexploded bomb, Wallace's letter was a live grenade. The mechanism he proposed for species change was closely similar to Darwin's. As Wallace requested, Darwin sent the paper on to Charles Lyell – bitterly regretting that he had not published his ideas earlier, as Lyell had advised him to. It was an agonizing time: two of the Darwin children were seriously ill and now his life's work seemed in ruins.

Lyell and Hooker between them arranged for Wallace's paper to be read at the Linnean Society, but only after two selections from Darwin's earlier work (extracts from the 1844 essay and part of a draft chapter on 'Variation of Organic Beings in a State of Nature' from the unpublished manuscript of *Natural Selection*), together with a copy of a letter Darwin had sent to the American botanist Asa Gray in 1857, outlining his complete theory. Their stated motive was not to establish Darwin's priority – although that is certainly what they did – but to demonstrate that the proposed theory was grounded in 'a wide deduction from facts' and 'matured by years of reflection'.[9] The Darwins' 18-month-old son, Charles Waring, died on 28 June 1858; the Linnean Society meeting took place on 1 July. Neither Wallace nor Darwin was there – Wallace did not even know it was happening.

In the event, the papers got very little immediate attention, but publishing the 'Big Book' was now out of the question. Darwin wrote *Origin* in little over a year; it was published on 24 November 1859, and he collapsed exhausted. He described it as an abstract because that is what it was: a précis of the main arguments from *Natural Selection*, which now became a blueprint for a future publication programme. He also described *Origin* as 'one long argument' and it was intended to be a persuasive one that would reach as wide an audience as possible. He crammed in as much of his evidence as he could and prefaced it with a robust statement of his position, denying that species were fixed types, claiming natural selection as the main but not the only mechanism by which they changed to produce new species, and that the forces through which this happened in the past were continuing and would continue to act. He even thought this process was speeding up as populations grew and competition became more intense.

The book sold out unexpectedly fast and Darwin was already making corrections for a second edition before the end of the year; six English editions were published in Britain in his lifetime, together with editions in America and translations into 11 other languages. Adam Sedgwick was appalled, describing some parts as 'utterly false & grievously mischievous';[11] George Eliot thought it ill written and unlikely to become popular, but welcomed it as opening the way for debate. Despite this prediction, Darwin's ideas rapidly entered not only the intellectual sphere but also the popular consciousness. In 1861 the satirical magazine *Punch* published a cartoon of a gorilla with a sign asking 'Am I a Man and a Brother?' – a play on the well-known anti-slavery medallion issued by Wedgwood pottery showing a kneeling slave with the motto 'Am I not a Man and a Brother?' It was the first of many cartoons in satirical magazines around the world – Darwin himself collected quite a few. Richard Owen published a highly critical review, and the Bishop of Oxford, Samuel Wilberforce, spoke out against Darwin at a meeting of the British Association for the Advancement of Science in 1861.

But Darwin had powerful friends as well as powerful enemies. He was famously defended by the comparative anatomist Thomas Henry Huxley – also known as

'Darwin's Bulldog' – who, when Wilberforce asked whether he would rather have a monkey as a grandfather or a grandmother, replied better a monkey for an ancestor than a man who joked about such things.

Darwin was proposed three times in the following years for the Copley Medal of the Royal Society, its highest honour, and was finally awarded it in 1864 against strong opposition, and with *Origin* rather pointedly omitted from the award citation. In the same year a small group of his supporters, intent on promoting his ideas, set up a private dining club, the X Club. Its members went on to position themselves strategically throughout the scientific establishment.

▲ **This cartoon, one of the earliest to comment on the implications of *Origin* for human development, was a play on the well-known anti-slavery medallion showing a slave with the motto 'Am I not a Man and a Brother?' (*Punch*, 18 May 1861, p. 206, Cambridge University Library)**

What Darwin did next

When *Origin* came out, Darwin was only 50 years old and most of his publications still lay ahead. The rest of his life can be seen as a research and propaganda programme already laid out in the plan for the never-completed 'Big Book'. It took him another 14 years to publish the major books, all of which were to some extent prefigured in the plans for *Natural Selection*.

He spent years experimenting on a range of plants to explore and explain the workings of natural selection, to rebut specific criticisms, and to hammer home the message of common descent. In 1871 and 1872 he published two books on the application of natural selection to humans (*The Descent of Man and Selection in Relation to Sex* and *The Expression of the Emotions in Man and Animals*), developing a supplementary theory of sexual selection and pushing forward the understanding of mind and emotion. He constantly widened his international network of collaborators, sending and receiving thousands of letters. He encouraged others to test his ideas and to do the experiments and make the observations that he could not. He helped promote careers and was frequently asked to lend his weight to various causes, which he almost always declined to do, at least publicly. He was sympathetic to Elizabeth Garrett's campaign to become the first qualified female doctor in Britain, but less so to Annie Besant and Charles Bradlaugh's attempts to make contraception legally available. Behind the scenes, he co-ordinated a robust defence of controlled vivisection, believing that science

▲ Charles Darwin on the terrace at Down House in 1878, photographed by his son Leonard (Cambridge University Library MS DAR 225: 1)

could not advance without it. He was a celebrity and, like any celebrity, had some fans he would rather avoid: he pleaded illness to many, and attempted to wriggle out of meeting a minor royal, the Duke of Teck. One admirer, the appropriately named Anthony Rich, insisted on leaving Darwin his fortune.

Like many older people, Darwin increasingly reflected on his own life and, in 1876, the year he became a grandfather, he started his *Recollections*, a rather idiosyncratic autobiographical account, continuing to add to it occasionally in the following years. With the help of his children, he took up genealogy, researching the life of one of his grandfathers, Erasmus Darwin, and helping the biographer of the other, Josiah Wedgwood I.

Darwin died on 19 April 1882 at Down House, where he and Emma had lived for 40 years. He had not been offered national honours during his life, but his scientific friends ensured that he was in death. Despite his own wish and the wish of his family that he be buried in the village where he had lived, he was interred with full public honours in the national church, Westminster Abbey, close to the grave of Isaac Newton.

2

Natural selection: the theory

'This preservation, during the battle for life, of varieties which possess any advantage in structure, constitution, or instinct, I have called Natural Selection.'

Charles Darwin, *The Variation of Animals and Plants under Domestication*, 1868

The full title of the book often referred to simply as *Origin* is *On the Origin of Species by Means of Natural Selection, or the Preservation of Favoured Races in the Struggle for Life.* (The word 'On' was dropped in the sixth edition of 1872, the last one to be published in Darwin's lifetime.) As a title, it gives almost the whole game away. Darwin proposed that new species emerge as a natural consequence of competition for resources between individuals, a competition in which small inherited advantages make the difference between life and death and, crucially, between living long enough to breed and pass on those advantages to the next generation and dying early and taking your less favourable attributes into oblivion with you.

He called this process 'natural selection' – similar in its effects to the artificial selection practised by plant and animal breeders that could so dramatically alter the appearance, health and habits of whole groups of organisms, from cattle to wheat.

Q: When is a species not a species? A: When it's a variety

Darwin argued that this simple fact of inherited variation, observable all around us, was sufficient to account for all the diversity of life. It was certainly observable all around Darwin: Britain's increasing prosperity was borne on the back of the great era of agricultural improvement through selective breeding, which was paying off in terms of higher-yielding crops, sturdier cattle, woollier sheep and ever faster racehorses. Variation within a species was obvious and accepted, but Darwin proposed that species themselves were nothing more than 'strongly marked varieties' and that, through the mechanism of natural selection, new species not only have been formed but also continue to be formed. Every adaptation to environment, every remarkable colour, texture, pattern, every mechanism for catching food and escaping from predators, could have emerged from a single primeval ancestor.

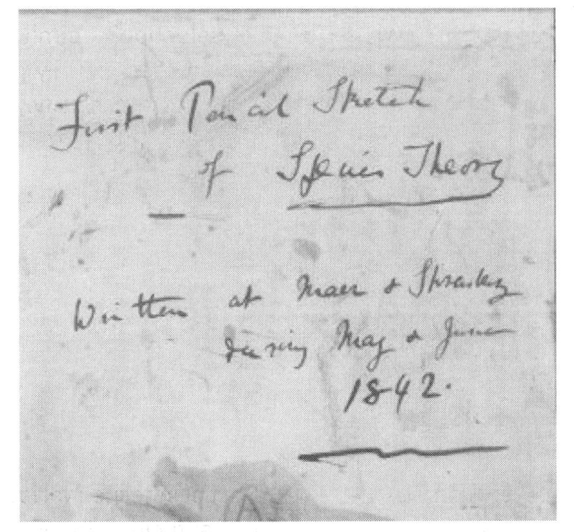
First Pencil Sketch
of Species Theory

Written at Maer & Shrewsbury
during May & June
1842.

▲ Darwin's note on the cover of the 'Pencil Sketch' – his first attempt at writing out his species theory (Cambridge University Library MS DAR 6: 1r)

Rather than thinking of the living world as the result of active adaptation to the environment – a purposeful crafting – Darwin saw it as the outcome of the extinction of those characteristics that do not work, or do not work as well. The beauty and marvel of the natural world were not deliberately built up like a painting or an intricate piece of pottery; it was more like a gorgeously weathered rock formation, just one of many possible latent forms left behind by the unthinking removal of extraneous material. Evolution is wasteful; extinction is as much a part of it as survival and change.

Natural selection vs survival of the fittest

The term 'natural selection' had its critics. The philosopher Herbert Spencer thought the analogy with animal breeding misleading because it suggested conscious action, and he proposed replacing it with 'survival of the fittest'. Alfred Russel Wallace went through his own copy of *Origin* crossing out every occurrence of 'natural selection' and replacing it with Spencer's term. But 'survival of the fittest' is often misinterpreted as 'survival of the strongest' (especially on management consultancy websites), which misses the point. Survival depends on being well adapted to your environment. That may mean being strong, but it may mean being small enough to hide, or light enough to run, or tall enough to reach fruit higher in the trees. And the environment itself is not static: the climate may change, your food source may dry up, or something else may arrive that eats what you eat – or eats you. The term 'adaptation' is also often misinterpreted as implying an active response by individuals (again, see management consultancy websites).

The last word

The very last word of *Origin* is 'evolved' and this is Darwin's first use of the word in print. He did not invent the term or the concept of evolution itself.

The word 'evolution' should come with a health warning. To 'evolve' is literally to 'unroll', but the term has done quite a bit of unrolling itself and has meant rather different things at different periods and in different contexts. It is often used today as shorthand for the process of species change through Darwinian natural selection – but, obviously enough, that is not what it meant before Darwin.[1] It was used from the late eighteenth century for any process of gradual change, especially from simple to complex, and was juxtaposed with 'revolution', implying sudden change. By the mid-nineteenth century it was used in biology to describe the development of varieties from multiple prototype forms, but it did not exclude the idea of a pre-existing plan for that development. It could, for example, be used to describe the development of an embryo, which was conceived of as a fully formed organism in miniature which had simply to blossom into its adult form. Charles Lyell had used the term 'evolution' to describe the progression from marine invertebrates to land creatures, and in 1852, seven years before Darwin, Herbert Spencer had already used 'evolution' for the transformation of one species into another.

What Darwin did was propose just how evolution – defined as the continual development of new species – could happen. It is a model that requires no outside agency or divine authority to keep it running.

It is not a teleological process – that is, it moves in no particular direction. There is no plan.

The species question

Darwin was not the only one asking the 'species question'. In 1860 the vast majority of naturalists, according to Darwin, still believed in separately created and unchanging species. By 1869 he was able to say in the introduction to the fifth edition of *Origin* that things were different. But there was a delicate balance to be struck between claiming novelty for his ideas, with all the credit in the intellectual and wider world that that could bring, and positioning his theories as just one step in a respectable tradition of scientific thought. Darwin published his own account of the earlier history of evolutionary ideas in the revised introduction to the third edition of *Origin* in 1860, starting with the French biologist Jean-Baptiste Lamarck (1744–1829) and listing a dozen or so others, including his own grandfather, Erasmus Darwin.

Lamarck believed that new forms of life were spontaneously and continuously generated, each beginning as a lower organism and progressing through the same stages to become successively higher ones. He proposed that organisms are transformed partly by their own efforts to respond to their environment and that these 'acquired characteristics' can be passed on to the next generation. The example Darwin used to explain Lamarck's theory is the idea that giraffes must have developed long necks through countless years of stretching up into the trees.

Alfred Russel Wallace (1823–1913)

In his first letter to Alfred Russel Wallace that survives, dated 1857, Darwin wrote: 'I can plainly see that we have thought much alike & to a certain extent have come to similar conclusions.'

Wallace was a Welsh land surveyor and teacher. From 1848 he earned a living collecting natural history specimens in Brazil and then the Malay Archipelago. In 1855, based on his observations of both living and fossil plants and animals, he proposed a branching model for the relationships between different species, and argued that 'Every species has come into existence coincident both in time and space with a pre-existing closely allied species'.[2] By 1858 he had worked out a way in which species could be transformed, very like the one Darwin was already calling 'natural selection'. Each knew the other was interested in species change and intended to publish, and they had swapped observations.

It has been argued that Wallace was cheated out of proper recognition for his work, but Wallace himself never said or appears to have thought this. The two men became lifelong friends, exchanging hundreds of letters discussing the details of their continuing research. They did not always agree, in particular about human evolution. Wallace was never well off, and his later belief in spiritualism damaged his scientific career. Darwin successfully petitioned for a government pension for Wallace, writing a personal letter to the Prime Minister, William Gladstone, and getting a personal reply.

Apart from Alfred Russel Wallace, the man who had most closely anticipated Darwin's own mechanism was a Scottish farmer, Patrick Matthew, but as he had outlined his ideas in only a few paragraphs as an appendix to an 1831 essay on naval timber,[3] it is hardly surprising that Darwin, along with everyone else, had overlooked it.

The origin of life, and the origins of *Origin*

Darwin never claimed that his theory could account for the origin of life itself, only for how different forms of life could emerge, and emerge moreover from a single shared ancestor. Although he left room in *Origin* for the idea that there might have been several ancestral forms, his later research confirmed his belief that there need only have been one.

> *'Will you honestly tell me (& I should really be much obliged) whether you believe that the shape of my nose (eheu) was "ordained & guided by an intelligent cause".'*
>
> Letter to Charles Lyell, 21 August 1861

The basic ingredients of Darwin's theory are: the natural tendency of populations to expand, resulting in competition for finite resources; the equally natural occurrence of small variations between individuals (differently shaped noses, for example); the heritability of those characteristics, not just from parent to child but within families (the shape of your nose may resemble that of your great-grandmother rather than that of either your father or mother); and, finally, the huge amount of time required for any slight advantage that that particular nose might give in competing for resources, for those of your descendants who inherit it to survive disproportionately generation after generation, and for noses of that shape to become a defining characteristic of the population.

Darwin was sensitive about the shape of his nose – he thought it was too big. He was also acutely aware of family likenesses. The number of marriages between his father's and his mother's families led to a standing joke that the Darwins were more Wedgwood than the Wedgwoods. Back in Britain after the *Beagle* voyage, in addition to getting married and starting a family (more Darwin–Wedgwood noses), he studied variations in natural populations and the power of selective breeding to change domesticated ones. The changes that pigeon fanciers were able to bring about, coupled with the occasional resurfacing of an older characteristic, struck him forcefully.

During his London years, when he was not talking pigeons in the pub or dogs and horses in the gentlemen's clubs, Darwin was reading. In 1838 he found what he needed: Thomas Malthus's *An Essay on the Principle of Population* (first published in 1798), from which Darwin drew the argument that populations will always increase exponentially until checked by lack of resources. Although Darwin was already thinking of species change in terms of a struggle to survive, his interpretation of Malthus's work made that struggle inevitable. And Darwin did not just read it: he was living it. The population of Great Britain doubled between 1815 and 1871; fluctuations in grain prices, both through bad harvests and as a result of protectionist laws, led to starvation and unrest, and in 1845, only a year after Darwin first wrote out the major points of his theory, the Irish Potato Famine provided a graphic illustration of the devastation that a change in the environment and suddenly increased competition for resources could wreak on a population.

The problem of time

Despite what pigeon breeders could do in a few generations, it was evident that, for natural selection to have accomplished everything that Darwin was proposing, it needed time, and lots of it. There was no consensus on the age of the Earth, or agreement on how to calculate it. Darwin followed Charles Lyell in believing that geological features have been created gradually through the action of familiar forces, and the dramatic evidence he had seen in South America of radical changes in the landscape convinced him that the Earth was far older than the generally accepted range of a few thousand years. He stuck his neck out, uncharacteristically, in *Origin* by publishing some calculations he had made just two years earlier in 1857 based on the rate of erosion in a nearby area of south-east England called the Weald.

His figure of a minimum of 300 million years for the age of the Earth was immediately challenged; he reduced it in the second edition of Origin to 100–150 million years and simply dropped it altogether after that. Despite some support for Darwin's timescales from other geologists, prevailing opinion settled

on a maximum of 100 million years based on calculations by the physicist William Thomson (later Lord Kelvin) using the assumed rate at which the Earth had cooled.

Today, the generally accepted age of the Earth is 4.5 billion years. Having that figure would have made Darwin's life a lot easier: having to allow for a much younger Earth made it far more difficult to account for the power of minute variations to create new species, and to explain the apparent lack of intermediate forms.

The missing link – and why it is missing

If every living thing is the result of processes of gradual change, why are we able to define species at all? Why do we not simply have a world populated with an amorphous continuum of unclassifiable individuals? If humans and baboons – and barnacles and crabs, and horses and hippos – share a common ancestor, where are the cousins that lie between?

Part of the answer Darwin gave lay in the incompleteness of the fossil record – the traces of many intermediate forms simply do not survive. But he also pointed out that in fact we do see gradation, that no two individuals *are* alike. He worked out that natural selection favours the extreme forms of any continuum, and that the middle ground – the links – would inevitably be the ones that went missing. This was the last piece of his evolutionary theory to be slotted into place, and he called it 'the principle of divergence'. In any competition for survival, forms that can exploit non-overlapping resources will be able to co-exist and thrive, whereas those whose needs are alike will struggle. Imagine a game of musical chairs where the chairs are a range of different heights and widths: there would be some at either end of the spectrum that only the very tall players or the very small players could occupy, so they could co-exist, leaving the middling-height players to fight it out. This means that, for any given environment, diversity will tend to increase.

Those finches ...

One group of related organisms in particular has come to symbolize Darwinian evolution – the several different forms of finches (*Geospizae*) each inhabiting a different island in the Galápagos archipelago. They are distinguished by differently shaped beaks adapted to exploiting different foodstuffs. One of the most persistent myths about Darwin is that the idea of natural selection came to him on first seeing those finches. Darwin not only failed to recognize their significance when he collected specimens in the Galápagos, but, because he did not keep the ones from different islands separate, he could not even use them to support his theories in retrospect; they are not mentioned in *Origin*. A now-famous illustration of the different finches appeared in his earlier *Journal of Researches* but, although he pointed out the gradation in their beaks, Darwin made no connection between that and the birds' geographical distribution.

‘Such facts would undermine the stability of species’

‘Here is my answering long shot about the cream-jug-nose: I should believe it to have been designed ... until I saw a way of its being formed without design, & at the same time saw in its whole structure ... evidence, of its having been produced in a quite distinct manner, i.e. by descent from another cream-jug whose nose subserved, perhaps, some quite distinct use.’

Letter to Asa Gray (after 11 October 1861)

In Darwin’s time most biologists believed that different species had emerged independently in separate centres of creation and spread from there, each designed for its own distinct habitat. But Darwin was bothered by the difficulty under this view of explaining, for example, the resemblance of living armadillos to the huge fossil mammal *Glyptodon* that he found in the same area of Brazil, or the overlapping territories of two species of the ostrich-like South American rhea. Travelling through

Patagonia, Darwin encountered plenty of one species of the rhea but only heard rumours of the second, rarer species until one evening in January 1834 when he realized, too late to save it all, that the *Beagle* crew were eating one for dinner. He grabbed as much of the carcass as he could, which cannot have gone down well with a boatload of hungry sailors.

Darwin made a better job of observing and systematically collecting different forms of another bird from four of the Galápagos Islands: mockingbirds (*Mimus*). It was thinking later about these, rather than the finches, in conjunction with different types of tortoise from the different islands, the two South American rheas, differences between foxes from East and West Falkland, and the armadillos and *Glyptodon*, that led him first to speculate, probably in the summer of 1836, that, if these really were more than just varieties, 'such facts undermine the stability of species'. But he was still unsure, and altered his note to read more tentatively that they 'would undermine the stability of species'.[4]

The ants and the bees: the problem with (no) sex

Social insects threw up a number of different problems for Darwin. One particular difficulty at first seemed fatal to his whole theory: the existence of not just one but several highly specialized neuter forms – nurses, builders and soldiers – that differed in appearance and behaviour not just from the males and females of the species but also from each other. How could their characteristics be passed on to the next generation if they themselves had no descendants?

Darwin had already demonstrated that there are natural variations in instinct, as in any other characteristic, which could therefore be modified through natural selection. He had also proposed a principle of correlation whereby a beneficial variation might naturally be linked to another variation with no benefit. The useless variation would still increase as a by-product of the useful one: it was along for the ride. In this case, sterility was the useless variation, and the instincts to build, fight or nurture the colony's offspring were the useful ones.

He still needed a way to explain how both variations are passed on. His answer was selection operating at family level. It goes back to those noses. If characteristics are common not just in direct lines of descent but also across families, colonies of related insects that tended to produce more of the useful neuters would thrive at the expense of those that did not.

The paradox of sterility

Thomas Huxley thought that natural selection could not be demonstrated until it could be shown to give rise to 'physiological species', by which he meant offspring of a common ancestor that could not interbreed. But in order to operate, natural selection depends on exploiting variations, and variations are more likely to occur when there is cross-breeding. Indeed, much of Darwin's work demonstrated that evolution favours mechanisms that *promote* interbreeding. So how could natural selection create species?

Darwin's answer to this apparent paradox is that mutual sterility is just a by-product of evolution. What we describe as separate species is merely what results when varieties have diverged in such a way that they happen to have become physically incapable of interbreeding. To put this in modern terms, humans cannot breed with other primates because the evolutionary changes that have allowed us to exploit different ecological niches have coincidentally produced physical changes that make our reproductive systems incompatible. Species are an evolutionary accident, but they seem such a fundamental distinction that we use them to classify living things. They are artificial categories we impose in order to make sense of the world.

The problem of progress

One of the biggest differences between Darwinian natural selection and earlier ideas about change in nature is that the Darwinian model is not inherently progressive, that is, moving towards a particular and more 'perfect' state. Although Darwin thought that natural selection would tend over time to create more complex forms from simpler ones, he continually reminded himself – not entirely successfully – to avoid hierarchical terms such as 'higher' and 'lower' in discussing living beings, and instead to use purely descriptive words like 'organized' and 'simple'. To be perfectly adapted to one's environment is a relative condition – there is no absolutely perfect state because there are no absolutely stable environments.

Mimicry

One of the most remarkable features of the natural world, and one which seemed to be evidence not only of design in nature but design by an intelligence with a rather whimsical sense of humour, was mimicry – the uncanny resemblance between separate species, particularly insects, living in the same area. Straightforward camouflage could easily be explained by natural selection, whether this was for defence (like the folded wings of a moth resembling the eyes and beak of an owl) or for some other purpose (such as carnivorous plants that attract insect prey by looking and even smelling like pieces of meat). But mimicry was a puzzle.

Darwin was delighted when two of his supporters came up with two different explanations for the evolution of mimicry, both based on natural selection. Alfred Russel Wallace's close friend and fellow explorer Henry Walter Bates concluded that some harmless butterflies had come to resemble other neighbouring species that had natural defences against the same predator – such as being poisonous. Slight variations that made individuals from the first species look a bit like the second had given them an advantage and had spread through the population and been increasingly refined until the appearance of the two species converged. Fritz Müller saw a different angle: he argued that different but equally protected species were also statistically likely to converge in appearance because the predators were more easily educated to recognize and avoid a single set of danger signals.

3

Science in practice

'... unbounded patience in long reflecting over any subject – industry in observing and collecting facts – and a fair share of invention.'

Charles Darwin, *Recollections*, 1876

It is impossible to discuss Darwin's ideas without saying something about how he did what he did. At least, it should be, but Darwin is so often talked about as though he were a theorist working entirely in the abstract – a philosopher rather than a practical scientist – that it is well worth underlining a few things about his working method.

Domestic science, Darwin-style

Darwin worked at home. The only time he ever had a workspace that he described as a 'lab' was when he was 12 years old and doing chemistry experiments with his brother in a shed in the garden. At Down, the animals and plants that fed and amused the household – the chickens, dogs, bees, pigeons, runner beans and hothouse plants – were also observed, measured and experimented on. Milk, eggs and meat, and even olive oil and wine, from the kitchen were appropriated to stimulate movement and digestion in carnivorous plants. Far from being confined to his study, Darwin's research invaded every area: there were twining plants on the parlour windowsill, insectivorous plants in the pantry and worms on the piano. The cock was fed various seeds (and his excrement carefully observed) in a study of the possible role of birds in plant migration.

'Here is a bee!'

Darwin used anything that came to hand, including his children. In notes for an unpublished study of instinct, he described positioning them at points in the fields around the house where he had seen bees pause along what he suspected were fixed flight paths. The children had to call out whenever they saw a bee at one of these 'buzzing places', and his hypothesis was confirmed as he heard the shout of 'Here is a bee!' taken up by each child in turn along the route.[1] Knitting needles were the essential equipment in another study of worm behaviour. Darwin had a small army, largely of women, poking knitting needles into worm burrows on hillsides all over the world to determine their angle.

But if this makes his methods sound antiquated it is misleading. A fun day out with Dad was also a significant contribution to the understanding of instinct and communication, and Darwin's work on the part worms play in the development of landscape was pioneering. Conducting science at home was not unusual at the time, or impractical for someone of Darwin's class. He could afford state-of-the-art equipment when it mattered: his Petri dishes were covered with bits of broken crockery and his dissecting table was a board on his study windowsill, but his complex of climate-controlled hothouses was specially designed and his microscopes were the best available.[2]

Darwin was well informed about the latest techniques and he made sure that he knew the leaders in emerging

▲ **Darwin's state-of-the-art hothouse complex, built for experiments on exotic plants (*Century Illustrated Monthly Magazine*, January 1883)**

fields. He may have studied peas and beans from the kitchen garden, but when he needed exotic plants he could call on two successive directors of the most prestigious botanical institution in the world, the Royal Botanic Gardens at Kew, to send him specimens. And as his research questions moved beyond what could be addressed in a domestic setting, he recognized these limitations and tapped into other resources. Men (and several women) all over the world did experiments suggested by Darwin that furthered his research agenda. He was effectively running a distributed research group.

When his research on digestion in plants required an understanding of their biochemistry, he sought out experts with access to laboratories, and encouraged them to do the experiments he could not. At Darwin's suggestion, one of his correspondents sent seeds from Brazil, not just to Darwin but to another correspondent in Germany so that they could measure the effect of different environments on growth and fertility. And when his botanist son Francis was working in a German university laboratory, Darwin proposed a number of experiments for the research team there.

Beetles and bladders: collecting and preserving

Commercial specimen collecting in the nineteenth century was big business. In the era of the *Beagle* voyage, Darwin was just one of many sending home phenomenal quantities of both organic and inorganic specimens from South America. He joked that there were more collectors in South America than carpenters or shoemakers or 'any other honest trade'.[3]

Darwin was a natural: he began as a child with rocks and shells and bits of pottery, and by the time he was a student his beetle collecting had reached the levels of competitive sport. At university he came up with strategies to get better collections than his friends: he paid labourers to strip the bark from trees and bag up debris from the barges that brought thatching straw to the city so he could search through them for stray specimens. Out in the countryside near Cambridge one day, he had just grabbed up two rare beetles, one in each hand, when he saw a third, so he popped one of the first ones in his mouth, where it immediately squirted acid down his throat: he gagged, spat it out, and dropped the other two as well. It was a story he was still telling nearly 50 years later.

Simply collecting specimens was not good enough. They had to be usable. This was an age before refrigeration. Darwin learned how to stuff animals while he was a student in Edinburgh, but still the first batch of mice he sent home from South America went mouldy – which cannot have been fun for Henslow, who was the one unpacking them – and other samples were unrecognizable by the time they arrived in England. Carelessly positioned labels got detached, and Darwin did not know at first how to handle the new types of plants he was encountering. Henslow sent detailed instructions on how to press large tropical leaves so they would fit on standard paper, and told him to fold part back to show both surfaces.

Darwin made use of whatever came to hand: old pill boxes and empty sauce bottles from the *Beagle* dinner table for small fossils, soil and shells; and empty barrels filled with brine or alcohol for larger creatures. His letters from the *Beagle* read like a correspondence course in contemporary preservation techniques, and he kept a running list of tips and tricks: to seal jars he used a layer of putrid bladder, then two layers of lead or tin foil, then another layer of bladder which he varnished over.[4] Crabs were not to be put in the barrels with other specimens because they turned the preserving fluid black, and all parasites were to be carefully removed from bird skins or nothing would be left but 'feathers and beaks'.[5]

Darwin's later work relied heavily on specimens sent to him by others and he in turn gave precise instructions on how to pack them. '... if any other species of Thalia shd. flower with you,' he wrote, 'for the love of Heaven & all the Saints, send me a few in tin-box with damp moss', signing himself 'Your insane friend, Ch. Darwin'.[6] Tin boxes, not wood, would prevent plants drying out. Shipping live plants any distance was almost impossible at the time of the *Beagle* voyage, but shortly afterwards a London doctor, Nathaniel Bagshaw Ward, invented a miniature portable glasshouse. One of the first people to use it was Darwin's close friend Joseph Hooker, and Darwin's later work on adaptation in orchids would have been impossible without exotic specimens sent in Wardian cases to Hooker at Kew.

Darwin and citizen science

Darwin was an avid collector of information as well as things, and after returning from the *Beagle* voyage he was increasingly reliant on others not just for specimens but also for observations. He was also quick to realize the potential of the nineteenth-century equivalent of crowd-sourcing: he approached the army about gathering information on resistance to disease through military surgeons, and tried to get a question on cousin-marriage included in the 1871 national census. Those attempts were unsuccessful, but he used the letters pages of cheap periodicals to publish questions directed at gardeners, nurserymen and farmers, and they responded enthusiastically.

'Dear Mr Darwin'

The indexes to Darwin's books read like magical mystery tours – silkworms from India, epilepsy in guinea pigs, feral cattle in the Falklands – and much of this information came to Darwin in the mail. His correspondents numbered many hundreds of men and women from a surprisingly wide range of backgrounds who provided information and research material, and most of whom he never met. The complete texts of all Darwin's more than 15,000 surviving letters make fascinating reading. They are being published in a print edition[7] and you can also read and search them online at www.darwinproject.ac.uk.

By his own admission, Darwin pestered people with letters. For example, Mary Barber, a settler and diamond prospector in South Africa who completely got the point of *Origin*, sent Darwin observations of camouflaged stone grasshoppers, perfectly adapted to small outcrops of rock (the only problem was that their very perfection made Darwin initially doubt her). Robert Swinhoe, a diplomat in China and an ornithologist, sent Darwin not only bird carcasses but observations on Chinese customs. John Scott, who worked in the Royal Botanic Garden in Edinburgh, spent so much of his time doing experiments for Darwin that he lost his job. (Darwin helped get him another in the Botanic Garden in Calcutta, where, conveniently, Scott was able to make new observations and experiments on Darwin's behalf.)

He perfected the use of standardized questions on a variety of subjects sent out through this ever-growing network. This was a technique he first used in 1839 when he distributed rather lengthy handwritten questions on animal (and human) breeding. It was an idea based on lists of questions for travellers sometimes put out by museums or learned societies. By 1868 he had honed his approach and a one-page printed questionnaire on emotional expression got responses from as far away as South Africa, Australia and China on 17 questions ranging from 'As a sign to keep silent, is a gentle hiss uttered?' to 'Is laughter ever carried to such an extreme as to bring tears into the eyes?' The number of responses – 66 – looks statistically insignificant to modern eyes but Darwin believed in quality over quantity, and stressed the importance of direct observation over reliance on memory.

Microscopes and photographs

Darwin was nothing if not thorough. For eight years, from 1846 until 1854, he worked almost exclusively on a taxonomic study of barnacles – so exclusively that his children thought studying barnacles was a universal male occupation, one asking a friend 'Where does your father do his barnacles?'[8] This was partly a training exercise. Darwin was perfecting the arts of microscopy and dissection alongside proving his credentials as a serious biologist. Creating taxonomies – the systematic classification of organisms into family groups – was considered the most important occupation for a scientist at the time.

Essential to this study was Darwin's microscope, and in this case he was not just an early adopter but also an innovator. He got advice on the best one to buy, but when this did not do quite what he wanted he modified the design. The manufacturer adopted his modifications and, in an early instance of celebrity endorsement, advertised the improved instrument as 'Mr Darwin's dissecting microscope'.

Artists recorded the landscapes of the *Beagle* voyage – Darwin was still somewhere in South America when the very first photographic images were being produced – but by the time *Origin* was published photography was all the rage. Darwin joined in the craze, swapping portrait photographs with his correspondents and friends, but he also built up a working collection, interested in the potential of this new technique for scientific research. Among the most striking are photographs of mental asylum patients sent to him by the doctor James Crichton Browne. Browne was interested in whether photography could help in diagnosis. Darwin was interested in capturing expressions to help establish just how emotion is conveyed and whether the same expressions are universal in humans. He hoped that photography would do this more accurately than an artist could.

Two of Darwin's sons took up amateur photography and, in addition to taking some of the most intimate portraits of their father, helped him use stereo photography to observe diurnal movement in plants, showing that they become dormant at night.

Technology and technique

'Mr Darwin brought in some photographs taken by a Frenchman, galvanizing certain muscles in an old man's face, to see if we read aright the expression that putting such muscles in play should produce ... And it came out at dinner, that several of us had been trying to move certain muscles before the glass!'

Jane Gray to Susan Loring, 1868[9]

Darwin was an innovator not just in technology but also in experimental technique. Between March and November 1868, he used visiting friends and relatives as guinea pigs in an experiment to test whether photography could accurately convey emotion. He showed them photographs from a book by the French physiologist Guillaume Duchenne, claiming to show individual emotions from surprise to deep grief, all stimulated by passing an electric current through particular facial muscles.

Darwin's records show that he refined his technique as he went along. The first few visitors were simply asked whether they agreed with the photographer's captions, but the later ones were shown only the images and had to use their own words to describe the expressions. Darwin recorded all the answers in rough tables, which he later analysed. This is essentially a single blind experiment. Similar experiments are still used in the study of artificial intelligence and conditions such as autism. Try for yourself at www.darwinproject.ac.uk/emotion-experiment.

▲ One of the photographs shown to visitors in 1868 – this one is supposed to show 'terror'. Of the 24 people Darwin showed it to, 20 described the expression as intense fright or horror, three as pain and one as extreme discomfort. (G. B. A. Duchenne, *Mécanisme de la physionomie humaine,* fig. 61 (Paris: Ve. Jules Renouard, 1862))

'The Hawks have behaved like gentlemen & have cast up pellets with lots of seeds in them; & I have just had a parcel of partridge feet well caked with mud!!!'

Letter to J. D. Hooker, 19 October 1856

Many of the experiments Darwin conducted were completely new and he had to work out the method as he went along. Others were quite simple but no one had thought to do them before. To show how plants might spread from one area to another, he extracted seeds from mud washed off birds' feet and from the pellets thrown up by birds of prey, and proved in both cases that they could still grow; he raised 54 plants from seeds stuck to a scab on the leg of a wounded partridge. It was also widely assumed that seawater destroyed seeds, making it close to impossible for plants to migrate from one remote land mass to another as a result of seeds floating over the ocean, but Darwin tested the effects of salt water experimentally and proved that many seeds survived.

He patiently followed the offspring of crossing experiments with plants through many generations, and made close observations of others as he studied their sensitivity or digestive ability. For months at a time in the 1860s and 1870s, we know exactly what Darwin was doing each day, sometimes each hour, as he went to check on his plants in the hothouse, the garden or on his windowsill, keeping careful experiment notes as they grew, flowered (or failed to), twined, climbed, exuded secretions around small lumps of meat (or failed to exude them round equally small pieces of glass), folded up their leaves in the dark and unfolded them in the light, thrived, or died.

Working sometimes for as long as 11 years on a single cross-breeding experiment, he developed painstaking methods for preventing insect pollination, covering plants with netting of precise gauge, and patiently fertilizing them with either their own pollen or pollen from other plants, marking the differently treated plants with coloured threads. Using a feather or a very fine bristle, he mimicked the action of insect pollinators in orchids, learning how the different plant structures ensured that pollen from one flower would be transferred to another while minimizing the risk of self-fertilization.

For the later work establishing that carnivorous plants really do digest organic matter, he kept meticulous records of their reactions to different substances, from fragments of coal to cooked egg. Investigating how some plants move to twine or climb or trap prey, he tried stimulating them with drugs ordered from his chemist (it is difficult to know sometimes whether what he was ordering was for his stomach or his research) and encouraged his university friends to try the same experiments with harder-to-acquire substances such as cobra poison. An investigation of movement in roots involved fixing tiny pieces of card to them with varnish. Darwin described his experimental method in minute detail in his publications, both to protect himself against criticism of his procedures and so that others might repeat the experiments.

4

Darwin as author

'I am turned a complete scribbler ... I never look further ahead than two or three Chapters – for my life is now measured by volume, chapters & sheets & has little to do with the sun ...'

Letter to C. T. Whitley, 8 May 1838

Reading and writing were part of a single process of knowledge creation, and Darwin took both very seriously. Books, 'those most valuable of all valuable things',[1] were the information highway of Darwin's age. They were both how you learned and how you influenced others.

Darwin was as systematic about reading as he was about experimenting. For several years he kept lists of books he wanted to read and ticked them off as he finished them. Books were tools for research, and he was completely unsentimental about them: he wrote all over them, and if they were too heavy, he cut them in half down the spine. He lent and borrowed them, exchanging recommendations with other researchers – he was part not just of a distributed research group but also of a distributed book club. He devised an information retrieval system, listing his marginal notes and the page numbers inside the back cover of each book. As he started a new research topic he scanned through the lists and made a master index of relevant books. If the book he was reading wasn't his, he made an abstract and filed it in one of more than 30 numbered portfolios on different subjects that he kept on labelled shelves in the room that was study, laboratory and library. As he wrote up his various publications, he crossed through the abstracts and notes as he used them with different-coloured crayons. When he thought of a new research question he started a new portfolio, sometimes shuffling notes from one to another, or cutting up and distributing letters that dealt with more than one subject.

The day job

Writing was the only paid work Darwin ever did. He kept track of the time he spent on each book, and followed his sales figures (and his earnings) keenly. He was a hands-on author who worked closely with his publisher, John Murray, and also with the printing firms Murray subcontracted. He decided on format and looked at samples of type. To cram in as much of the evidence as possible without deterring the more casual reader, a couple of his books used two sizes of type, one for the main narrative and a smaller one for the ... small print. Darwin organized the illustrations himself, rejecting any inaccurate engravings, and in the 1860s researched new methods of cheap photographic reproduction. *Expression* was the first commercial English science book to use photographs, and it was Darwin who persuaded his reluctant publisher to commission them from a company pioneering the new technique of heliotype.

Darwin as travel writer

'... consider the infinite importance to a young author of his first proof sheets.'[2]

By the time he came back from the *Beagle* voyage, Darwin knew he wanted to be a writer and was excited by the idea. His first book was a travel book based on his diary, which he had written with an eye to possible publication. It was part of a tradition of scientific travel writing of the kind he enjoyed

reading himself. Now known as *The Voyage of the Beagle*, it was first published as part of FitzRoy's three-volume *Narrative of the Voyages of HM Ships Adventure and Beagle* but was quickly reissued as a standalone book under the unpromising title *Journal of Researches into the Geology and Natural History of the Various Countries Visited by HMS Beagle*. It was a success; Darwin sold the copyright and the second edition appeared in 1845 as *A Naturalist's Voyage around the World*. It remains the second-most widely read of his books after *Origin*.

Darwin's sisters teased him about basing the writing style of the *Journal of Researches* on that of his literary and scientific hero, Alexander von Humboldt. Humboldt's account of his own expedition to South America was one of Darwin's most prized possessions on board the *Beagle* and he kept it to the end of his life. On board, Darwin's hammock hung beside the ship's library – a pooled collection of more than 300 volumes divided between scientific manuals and travel writing, with a couple of dictionaries and, of course, the Bible, another book that occasionally influenced Darwin's use of language.

'You ask about my book & all that I can say is that I am ready to commit suicide ... I begin to think that everyone who publishes a book is a fool.'

Darwin on writing *Insectivorous Plants*, in a letter to J. D. Hooker, 10 February 1875

Darwin thought a great deal about how to write but it remained hard work for him and he thought his style 'wretched'. He always planned the structure of his books carefully, but gave up trying to write straight off in well-formed sentences, and instead dashed off a first draft as quickly as possible and then rewrote it. Few of the original manuscript pages of his books survive, but those that do show evidence of a good deal of reworking, and not only by Darwin. The earlier books, in particular *Origin*, were read by Hooker, Georgina Tollet (a family friend) and his wife. Later books were largely read and edited by Emma and the children. Henrietta helped edit *The Descent of Man*, and George checked the proofs for the second edition; Francis became his father's official secretary from 1874 and worked closely with him on the botanical books.

Darwin deliberately targeted the general reader. Reaching a wide audience was part of a conscious strategy to promote acceptance of his theories. 'I ... try', he explained in a letter dated 1863, 'to get subject clear as I can in my own head, & express it in the commonest language which occurs to me.'[3] He avoided introducing new scientific terms ('natural selection' and 'pangenesis' are pretty much the only ones), he published relatively few papers in academic journals, and his work was rarely heard first in scientific meetings. He haggled with his publisher over the cover price of his books, keeping them as cheap as possible to boost circulation, and insisted that his last edition of *Origin* should be a mass-market one. He encouraged foreign translations, working with the translators to get the terminology right.

'I am, as it were, reading the "Origin" for the first time, for I am correcting for a 2nd. French Edition; & upon my life, my dear fellow, it is a very good book.'

Letter to J. D. Hooker, 10 April 1865

The wider scientific world first met Darwin through some of his letters from the *Beagle*, which Henslow had printed and circulated through the Cambridge Philosophical Society. Books and letters weave in and out of one another: letters comment on books and are quoted in print in their turn; books of all kinds are recommended and discussed; and borrowing and lending are negotiated. Darwin's books are intimate and conversational, often written in the first person, and very reminiscent of his letters, which, together with his diary, are where he first honed his writing skills. He leads his readers on shared adventures of discovery, inviting us to tackle doubts and difficulties with him. Successive editions of his books, in particular of *Origin*, carried on a dialogue with his critics, introducing new evidence, and extending or even altering arguments.

The language and structure of *Origin* owe a great deal to the influence of William Paley's widely read books promoting divine design in nature, *A View of the Evidences of Christianity* (1794) and *Principles of Moral and Political Philosophy* (1785), both of which were required reading for Darwin as a student at Cambridge and would have been very familiar to Darwin's readers. Darwin admired Paley's writing style: he employed the same literary devices and took the same kinds of evidence as Paley – detailed observations of the natural world – but turned them to exactly contrary ends.

'What is the good of having a friend, if one may not boast to him? I heard yesterday that Murray has sold in a week the whole edit. of 1,500 copies of my book.'

Darwin on the sale of *Variation*, in a letter to J. D. Hooker, 10 February 1868

Although the print runs of Darwin's books (generally between one and three thousand per issue) sound small by modern popular-press standards (they are still quite enviable in academic terms), *Origin*'s first-year sales of around 4,000 stand up pretty well alongside the 20,000 for Samuel Smiles's *Self-Help*, one of the most successful of all Victorian books (and coincidentally published on the same day). By 1876, 16,000 copies of *Origin* had been sold in England, but this does not tell the whole story. A high proportion of Darwin's books went to libraries and were lent over and over again; they were reviewed in depth and their arguments widely reported in the press; and although Darwin himself did not give popular lectures about his work, he cultivated those, like Huxley, who did. When *Expression*, which had been eagerly awaited by a public to whom Darwin was now a celebrity, sold out on the first day, Darwin's rather excitable editor considered calling in the police to protect the publishing house. And Darwin was both pleased and astonished when his last book, *Earthworms*, outsold even *Origin*.

▲ **Darwin thought children's expressions more natural and he had a number of photographs like this in his collection. He reproduced a similar one of this little boy in *Expression*, which was among the first books illustrated with photographs reproduced by the new and cheaper heliotype process. (Cambridge University Library MS DAR 53.1: 129)**

Darwin on (and in) fiction

Reading light fiction was a way of winding down from academic work. The novels that were serialized in weekly magazines and passed around friends and family were the equivalent of soap operas; their characters and plots were part of the shared common currency of the age. Darwin liked his novels to have pretty, likeable heroines, and hated unhappy endings, against which, he joked, a law should be passed. He made references to books by Jane Austen in his letters home from the *Beagle*; he read Wilkie Collins's *The Woman in White* to his children; his wife and daughters read the latest bestsellers by Charles Dickens and Elizabeth Gaskell aloud in the evenings to the rest of the family, and he and Joseph Hooker swapped reading suggestions and comments. Anthony Trollope was a favourite with all of them: Darwin particularly liked *Framley Parsonage*.

Darwin as a fan

Darwin hugely admired the controversial novelist George Eliot. Eliot and her common-law husband, George Lewes, were friendly with Darwin's brother, and Lewes had written a number of essays on Darwin's theories and had also reviewed his books. Darwin used his connections to invite himself to tea with the famous author (ironic considering how often he made excuses to avoid his own fans) and later begged further invitations on behalf of his wife and daughter.

In later life Darwin lamented that he had lost the taste for higher works of literature, in particular Shakespeare or any kind of poetry, although he remained fond of John Milton. Milton's *Paradise Lost* had been his favourite book as a young man; he had carried his copy in the jungles and mountains of South America, and quoted Milton more than once in his own lyrical descriptions of the grandeur of the Andes. He later quoted from both

Shakespeare's *Hamlet* (on 'grief'), and Dickens's *Oliver Twist* (on 'rage') in *Expression.*

Darwin influenced writers in his turn, his theories becoming part of popular culture. Charles Kingsley's *The Water Babies*, where the drowned chimney sweep Tom is reborn with gills and develops (physically and morally) into a gentleman, reconciled evolutionary theories with Christian principles in a sharp social satire; Samuel Butler's *Erewhon* described a world of machines evolving consciousness through Darwinian selection; and George Eliot was criticized for the degree to which she incorporated scientific ideas and language into her novels, in particular Darwinian natural selection into *Middlemarch* and *Daniel Deronda*. Elizabeth Gaskell, a distant relative of Darwin's, left an uncompleted novel, *Wives and Daughters*, with a character of a young scientist modelled on him.

In the decades after Darwin's death, one of Huxley's students, H. G. Wells, explored the idea that humans might evolve into two separate species in his most famous science-fiction story, *The Time Machine*, and Thomas Hardy incorporated ideas of competition, inheritance and extinction into *Tess of the D'Urbevilles.* There are now more than 17,000 books with the word 'Darwin' in the title.

5

Little notebooks, big ideas

'... my work on the species question has impressed me very forcibly with the importance of ... what people are pleased generally to call trifling facts. These are the facts, which make one understand the working ... of nature.'

Letter to Leonard Jenyns, 12 October 1844

In the last sentence of his last book, *The Formation of Vegetable Mould, through the Action of Worms* (generally known as *Earthworms*), published just months before he died, Darwin pointed out that worms, like the coral polyps in one of his first publications, were apparently insignificant, even primitive, creatures, whose accumulated impact over time could nevertheless change the environment. As long before as 1837 he had shown that worms were responsible for covering up surface features in the landscape by continually bringing up earth.

In much the same way, Darwin's world-changing ideas were built up from detailed observations and seemingly insignificant experimental results. And they also took time. Many of the theories that first appeared in outline in his notebooks of the 1830s and 1840s were not seen

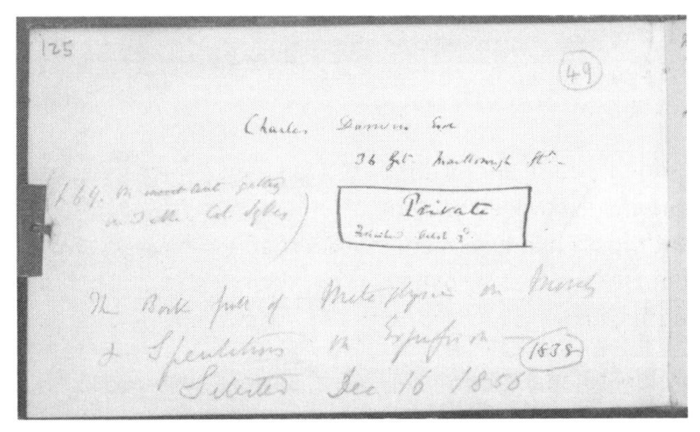

▲ **The inside cover of Darwin's notebook on 'Metaphysics, Morals, and Speculations on Expression', marked 'Private' (Notebook 'M', Cambridge University Library, MS DAR 125: ifc)**

in print for decades as he worked to amass the data that first tested and then supported them.

Seashells on mountains and forests in the sea

By the end of the *Beagle* voyage, Darwin had enough evidence to argue that many features in the landscape could be explained by the slow movement over millennia of vast blocks of the Earth's crust see-sawing on a shifting molten substructure, with elevation in one area matched by subsidence in another. The constituent parts of this idea were not new – Charles Lyell was championing a gradualist interpretation of the Earth's geology and Alexander von Humboldt had identified large-scale geological directionality (called loxodromism) – but Darwin put them together, provided evidence, and showed that his hypothesis could explain the otherwise inexplicable.

Eroding out of one of the highest passes in the Andes, Darwin discovered marine shells and fossilized trees that must once have been submerged in seawater. How could the trees have been carried from dry land, to ocean depths, and back up to the top of a mountain range?

A clue to the answer lay in a series of unfortunate events – events apparently unconnected but which Darwin realized were explicable as the results of a single seismic episode beneath the Earth's surface.

After being caught in an earthquake while they were surveying the coast of Chile, the *Beagle* crew repeated their earlier measurements and discovered that there was a small but permanent change in the relative height of land and sea. Here was confirmation of Lyell's gradualist interpretation of geological features. But shortly before the earthquake they had also witnessed the eruption of the distant inland volcano, Mt Osorno, and soon afterwards Darwin, travelling further up the coast, saw the devastation caused by a massive tidal wave. What if these were connected, the local effects symptomatic of a global event? Darwin literally pieced together his observations, gluing pieces of paper into long strips as he sketched geological cross-sections of almost the entire continent. Although differently tilted and distorted, the strata on either side of the Andes matched. The Earth moved. And it moved in large blocks.

Armed with this hypothesis of subsidence and uplift, Darwin went on to propose solutions to two geological puzzles. The first made his scientific reputation; the second he later described as 'one long gigantic blunder'.[1]

Coral reefs and the Great Glen

One of the many dangers faced by the crew of a sailing ship was oceanic coral atolls, barely visible above the waves, largely uncharted, and lethally destructive to wooden hulls. As coral polyps, the creatures that create the reefs, cannot live in deep water, the origin of the atolls was a mystery. Lyell had suggested that the colonies must be growing on the circular craters of undersea volcanoes. Darwin worked out, however, that they could have started life as reefs encircling islands – islands that had gradually sunk beneath the water as the large landmasses beneath them subsided. The coral had continued to grow as the land went down, keeping pace with the surface of the ocean generation on top of generation, layer on layer. Darwin's explanation for their formation fit every known form of coral reef, and the distribution of coral reefs around the world supported Darwin's theory of global-scale, tilting crustal blocks. He published the two conclusions together in 1837 and his simple and elegant theory won him wide admiration.

Darwin's coral reef theory was only proved experimentally in 1952 when a US Atomic Energy Commission borehole on Eniwetok near Bikini Atoll in the Pacific showed coral to a depth of 1,300 metres (4,158 ft) sitting directly on bedrock. The site was rediscovered in 1976 and a handwritten sign put up announcing that 'Darwin was right'.[2]

At first sight, the same explanation seemed to Darwin to account just as neatly for the so-called 'parallel roads' of Glen Roy – a series of striking horizontal terraces running round the valleys off the Great Glen in the highlands of Scotland. They appeared to be the remnants of ancient beaches but the glens were open, with no sign of any dam to keep in a lake. Darwin proposed that the glens had been periodically flooded with seawater, the levels fluctuating as the land mass itself rose and fell. Ignoring the slight inconvenience of a complete lack of marine fossils, and perhaps rushing a bit as he was about to get married, Darwin went into print shortly before Louis Agassiz announced his glacial theory and provided, in the form of large prehistoric bodies of ice, the elusive dams that made the lake theory far more tenable.

The sixties: all about sex

'You are quite right that there is nothing direct in my Book on the final cause or manner of origination of the sexes. It seems to me one of the profoundest mysteries in nature ... [T]he old saying was, "It was a wise child who knew who his Father was", but now the saying might be, "It was a wise child who knew whether he had a Father."'

Letter to C. G. B. Daubeny, 16 July 1860

By the time he wrote this, Darwin did have some ideas about the origins of the separate sexes. His eight-year study of barnacles had had some unexpected results. In contrast to most Crustacea, barnacles are largely hermaphroditic, but Darwin found a few sexual oddities: in one species there were also a few females, and both females and hermaphrodites had tiny parasitic rudimentary males. He called these 'complemental males', where the original hermaphroditic form was losing its male organs and a two-sex species was emerging. The question was why.

Variation drives evolution. Cross-breeding maximizes the possibilities of producing variations. Sex is one way to ensure cross-breeding. Much of Darwin's work in the 1860s was designed to explore and then to demonstrate the advantages of cross-breeding, to explain the complex mechanisms that promote it, and to show that the mechanisms not only promoted, but were themselves the products of, natural selection. Even hermaphroditic barnacles rarely self-fertilize – they make up for being stuck in one place by having enormously long penises (as Darwin was relieved to discover).

Three sexes are better than two

In 1861 Darwin published a paper on the two forms of primula, one with long anthers and short styles and the other with short anthers and long styles, demonstrating that significantly more seeds were produced by crossing the two forms than by breeding within one form, or by self-fertilization. This was another step on the way from hermaphroditism to two distinct sexual forms.

In 1862 he described an orchid, *Catasetum tridentatum*, which had male, female and hermaphrodite flowers on the same plant. And in 1864 he showed that purple loosestrife (*Lythrum salicaria*) is a fully formed three-sex species, capable of interbreeding in 18 different combinations.

No sex please, we're cleistogamic

What about reproduction without sex (or without sexual partners, at least)? Some species of water lily have flowers that remain not only closed (cleistogamic) during the whole reproductive cycle but under water, so that cross-fertilization seems impossible. Darwin had two answers to this: first, he pointed out that even a few offspring are better than none and that, although cross-breeding may produce more and better offspring, it carries risks – you may be unlucky in love. Self-fertilization at least means that you do not have to go looking for a partner (or rely on an insect to go looking for you). You will have fewer and less variable offspring when you reproduce, but you are pretty much guaranteed to produce something. Nature, and the water lily, hedges its bets.

However, that does not spell the end of evolution in water lilies. Darwin also argued that exclusive self-fertilization in sexual organisms is an illusion. They all cross-breed sometimes; it is just that some of them do it only very rarely. He discovered that even the usually cleistogamic water lilies produced the occasional open, above-water flower – just enough to mix it up a little.

Co-adaptation, or friends with benefits

Orchids were a perfect study for Darwin. They are both beautiful and beautifully varied and, to many contemporaries, seemed obvious examples of divine creation.

In 1862 Darwin published *Orchids* (full title: *On the Various Contrivances by which British and Foreign Orchids are Fertilized by insects, and on the Good Effects of Intercrossing*), in which he tackled the issue of separate creation of species head on. He demonstrated that, despite a vast amount of modification and specialization, all orchids share the same blueprint. Organs that fulfil very different functions in different species are nevertheless demonstrably adapted from the same basic pattern, and in other cases organs survive in some species but no longer serve any purpose.

Darwin was careful not to completely 'unweave the rainbow', however: he used the word 'beautiful' inordinately often and stressed that nature can still be appreciated even when it is understood as 'living machinery'. The most breathtaking mechanisms are adaptations that first attract insects and then induce them to pick up pollen from one plant, but allow them to deposit it only when they enter another.

The predicted moth

The co-adaptation of individual orchid species to individual insect pollinators was so marked that, when an orchid was discovered in Madagascar with an extremely long curled nectary, Darwin predicted that a moth with a similarly long proboscis must exist. The orchid is *Angraecum sesquipedale*, the Christmas Star Orchid, and its pollinator, exactly as Darwin had imagined it, was discovered in 1903.

Animal, vegetable, mineral?

'It has often been vaguely asserted', Darwin wrote in 1865, 'that plants are distinguished from animals by not having the power of movement', but, he continued, 'it should rather be said that plants acquire and deploy this power only when it is of some advantage to them.'[3] Plants generally do not need to move for either food or sex and, when they do move, Darwin showed that movement itself is an adaptation to allow exploitation of otherwise inaccessible resources.

By studying plants that twine and climb, Darwin was able to show that the boundaries – always his favourite area – between plant and animal are more blurred than we tend to think. Plants, especially when young, have some natural ability to move – just watch a stop-motion film of a seedling – and he found evidence that tendrils, although performing the same function in different plants, have been adapted in some cases from leaves and in others from flower stems. Not only could plants move, but they could also sleep – a mechanism for reducing heat loss through radiation – catch animal prey, and digest it.

He studied one plant, the sundew (*Drosera*), particularly closely. It was, he said in 1863, 'a wonderful plant, or rather a most sagacious animal'.[4] He demonstrated that sensation from one part of the leaf was transmitted to another through something resembling a nervous system, and that it not only trapped insects (and other bits and pieces) but digested them by producing acids very like those found in an animal stomach. It was all a way for organisms to exploit the full range of habitats, including those, like anaerobic bogs, where roots were no use for nutrition.

The man-eating tree of Madagascar

Carnivorous plants became all the rage once it was known that Darwin was working on them, with spoof stories of man-eating jungle plants making the rounds of the press. Darwin only realized that one description of a large tree exuding a sticky trail of sweet fluid was a hoax when he came to the part where it ate a woman.[5]

In two studies right at the end of his life, Darwin attacked the apparent gap between plants and animals from both directions. He resumed his research on worms, untouched since the 1830s, this time exploring their sensitivity to light, heat and cold. He demonstrated that, though they could not hear and had little sense of smell or taste, they were sensitive to vibrations and touch, and, like higher animals, were intelligent enough to build burrows and manipulate objects. Although they had no stomachs, he showed that worms, just like carnivorous plants, secreted an acid to aid digestion. At much the same time, working with his son Francis on a study of sensitivity in plant roots, he showed that a root could transmit sensation to produce movement in other parts of the plant. It acts, he said, 'like the brain of one of the lower animals'.[6]

The great god Pan

'Hypotheses may often be of service to science, when they involve a certain portion of incompleteness, and even of error. Under this point of view I venture to advance the hypothesis of Pangenesis.'

Charles Darwin, *The Variation of Animals and Plants under Domestication*, 1868

When he wrote *Origin*, Darwin had no explanation for how heredity worked. He just had to say that it was obvious it did. In 1868, at the back of an extremely dense two-volume work, *Variation of Animals and Plants under Domestication*, he tentatively sneaked out a 'provisional hypothesis' of heredity that he called 'pangenesis'. Although very few were convinced at the time, and Darwin himself never felt fully confident, it lives on in an indirect way in the modern word 'gene'. Pangenesis was intended to explain how characteristics could be inherited from parents, how they could skip generations, how they could be passed on intact or as intermediate forms, how reversions or variations could occur, and how, in some animals, amputated limbs could regrow.

Lord Morton's Law

An extra complication for Darwin in trying to explain how heredity worked was the apparent need to explain the transmission of the characteristics of one mate to the offspring of a subsequent mate. This was known as Lord Morton's Law and it was shown after Darwin's lifetime to be based on mistaken evidence.

Darwin concluded that all characteristics of both sexes and of previous generations must be latent in any given individual. Cell biology was in its infancy, but he postulated that the cells of the body, each capable of forming a given part, produced tinier particles, which he called gemmules, that continued to circulate in bodily fluids, each having the same ability to form a particular part of the body as the parent cell. The gemmules could be passed on in a dormant state down through the generations and aggregate to generate buds, eggs or sperm.

Although his cousin Francis Galton (best known for his later involvement in the eugenics movement) conducted some gruesome experiments on rabbits, the hypothesis was difficult to test experimentally, and even Darwin's nearest allies remained dubious. 'It seems that the poor infant Pangenesis will expire, unblessed & uncussed by the world', he wrote, 'but I have faith in a future & better world for the poor dear child!'[7] Huxley advised caution but did not want the responsibility of quashing the idea completely: 'Somebody rummaging among your papers half a century hence', he prophesied, 'will find Pangenesis & say "See this wonderful anticipation of our modern Theories – and that stupid ass, Huxley, prevented his publishing them"... I am not going to be made a horrid example of in that way.'[8] As no one came up with anything better in his lifetime (or not that he knew of, anyway), Darwin almost alone remained a believer in his 'great god Pan'.

6

Darwin and human evolution

'... man, like every other species, is descended from some pre-existing form ...'

Charles Darwin, *The Descent of Man*, 1871

Humans were a problem for Darwin. Exposure to the full range of the human species on the *Beagle* voyage forced him to confront the fact that they varied as much as any other organism, and that their behaviour did not necessarily distinguish them from other species of primate. Back in London, he spent some time observing the zoo's newly acquired orang-utan, Jenny, and at the same time acquired a new human specimen in his baby son, William. He started making notes on both, looking not for what made them different but what made them the same.

That natural selection must apply to humans as much as to any other animal was evident to him from the beginning, but Darwin made a strategic decision to leave humans out of *Origin* on the grounds that natural selection did not need to be any more controversial than it already was. Or he almost did: borrowing a trick from the serial novels of writers like Charles Dickens, Darwin ended the book with the cliffhanger that 'light will be thrown on mankind and his origins'. Anyone waiting for this to happen had a long wait. Darwin did not tackle human evolution head on in print for more than ten years until he published *The Descent of Man and Selection in Relation to Sex* in 1871 (hereafter referred to as *Descent*) and its companion volume *The Expression of the Emotions in Man and Animals* in 1872.

He rather hoped that the light would be thrown in the meantime by someone else, and to some extent it had been. Thomas Huxley, in a lecture to the Linnean Society in 1858 misleadingly entitled 'The distinctive characters of Man', had already argued before the publication of

Origin that, physiologically, humans are 'at one with the rest of the organic world' and in another lecture the same year had even asserted that not only the physical, but the 'mental and moral faculties' are 'essentially and fundamentally the same in kind in animals and ourselves'.[1] Huxley published his lectures on human evolution in *Evidences as to Man's Place in Nature* in 1863, and in the same year Charles Lyell published a book Darwin had high hopes for: *Geological Evidences of the*

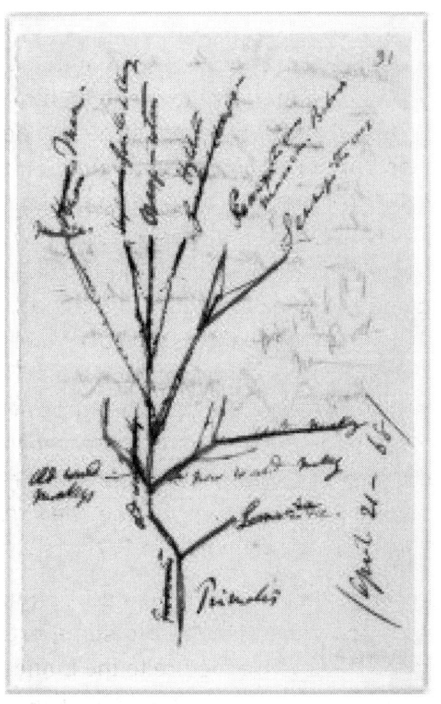

▲ **Darwin's diagram of the evolution of primates, 21 April 1868; 'Man' is at the top left (Cambridge University Library, MS DAR 80: B91r)**

Antiquity of Man, but to Darwin's bitter disappointment Lyell was publicly cautious about admitting change in any species, let alone in humans. He tried to get Alfred Russel Wallace to publish, offering Wallace his own notes, but Wallace also shied away from applying the full force of natural selection to humans.

There was a dilemma for Darwin in addressing questions about humans at all: the whole thrust of his argument was that there is no fundamental boundary between human and animal, that humans *are* animals, so that to single them out for special study ran the risk of putting them back on the very pedestal he was trying to knock them off. In his introduction to *Descent* Darwin therefore set out his stall: he was looking at humans not as a special case but as a representative in-depth study of a single species. He stressed the artificiality of looking at any species in isolation, and so he spent a good part of the book talking about our relationship to other species. There is as much about other primates – and birds and insects – as about humans, and *Descent* is as much about the ways in which other animals behave like people as about people behaving like animals.

By the time *Descent* was published, the ground was better prepared for what is actually a hard-hitting book. Darwin was a good strategist: get the theory itself in the public domain, get people used to it, put the clues in place for those prepared to follow them, and wait. Holding fire on *Descent* also gave time for the arguments against applying natural selection to humans to crystallize: by 1870 Darwin knew who his enemies were, and their firepower, and that allowed him to fine-tune his counter-attack.

Humans as animals

'Nothing is to be depended on but the great hippopotamus test.'

Charles Kingsley, *The Water Babies*, satirizing the scientific debate over the hippocampus, 1863

Bodies

Much of the apparent evidence for humans having a unique physiology had already been disposed of by the time Darwin wrote *Descent*. In 1862 the main pillar of the argument that human brains were different from animal brains was demolished in a dramatic bit of academic theatre. At a meeting of the British Association for the Advancement of Science, Thomas Huxley and William Henry Flower dissected an ape brain and produced a hippocampus, a small structure in the medial temporal lobe that helps create and store memory – a structure the palaeontologist (and by now implacable Darwin enemy) Richard Owen had claimed was exclusively human.

In the same year a French physiologist, Guillaume Duchenne, published a study contradicting the accepted belief that the human face had unique muscles designed to express exclusively human emotions. It was Duchenne's photographs that Darwin showed to his visitors to test how accurately they could identify emotion (see Chapter 3), and he used several of Duchenne's illustrations in his own book on expression.

Ernst Haeckel (1834–1919)

Haeckel was a German zoologist and an abrasive but very successful popularizer of Darwin's theories. He wrote two particularly influential books, *Generelle Morphologie der Organismen* (*General Morphology of Organisms*, 1866) and *Natürliche Schöpfungsgeschichte* (*The History of Creation*, 1868). He is best known for his theory that 'ontogeny recapitulates phylogeny', that is, that the developmental stages of embryos replay the developmental stages of the species as a whole, and for coining the word 'ecology'. Another of his statements, that 'politics is applied biology', was later notoriously adopted by Nazi propagandists.

In 1868 Haeckel published striking studies comparing the embryos of chicks, turtles, dogs and humans. His plates showed almost identical structures and stages of development across all four species, and though there was later controversy over the degree of licence in Haeckel's interpretation of what he saw, at the time the images were a powerful argument in favour of common ancestry. Haeckel's conclusions about the genealogy of humans agreed so closely with Darwin's own that Darwin said that, if he had seen the book any earlier, he would not have needed to write *Descent* at all.

> *'Almost thou persuadest me to have been "a hairy quadruped, of arboreal habits, furnished with a tail and pointed ears".'*

Asa Gray to Charles Darwin, 14 April 1871, on reading *The Descent of Man*

Darwin argued that the superficial differences that seemed to set humans apart from apes – our lack of body hair or a tail, and our ability to walk upright on two legs – could be accounted for perfectly well by natural selection. He also put forward some new evidence in *Descent* for shared physical features across species, identifying, for example, the small cartilaginous lump on the edge of the upper fold of the human ear as the vestigial remains of the pointed ears of our ancestors. He christened this the 'Woolnerian tip' after the sculptor Thomas Woolner, who had told Darwin about it during sittings for a marble bust. Try feeling your own ear; the little lump is more pronounced in some people than others.

Minds

> *'... can the mind of man, which has, as I fully believe, been developed from a mind as low as that possessed by the lowest animal, be trusted when it draws such grand conclusions?'*
>
> Charles Darwin, *Recollections*, 1876

It was far more difficult to demonstrate continuity between the mental characteristics of humans and other animals. At first sight we seem so clearly distinguished by our intelligence, our ability to reason and our unique use of articulate language. But even these, Darwin argued, were differences of degree and not of kind. He hunted out anecdotes of rational behaviour in birds, mules, snakes and even crustaceans, and gave his own account of apparent reasoning by his dog. He pointed to evidence of intelligence (which he believed to be inherited) in a range of animals, and to widely varying degrees of intelligence among humans, and argued that, like any inherited advantageous variation, intelligence could have been developed in humans through natural selection.

Far from being specially endowed with an ability for articulate language, Darwin pointed out that humans have to learn it. It is a tool that we become increasingly competent at using, and in this we are no different from birds, which have to learn to sing the same songs as their parents. What we are born with is an instinctive use of sound coupled with the instinct to imitate, and a desire to communicate, which is the basis of what Darwin described as 'the half-art and half-instinct of language' (*Descent*).

Darwin, and others, argued that human language evolved from this instinctive use of sound through something rather like natural selection, with relationships clearly traceable between now-distinct languages, some becoming dominant, others dying out. But this is also true in animals: dogs use sound to communicate differently from wolves, and birds have the equivalent of regional dialects. Our greater development of language, Darwin thought, had further strengthened our intellectual abilities through a sort of feedback loop, allowing us to develop more abstract ways of thinking.

Souls (consciousness and conscience)

Even self-awareness and the ability to speculate on the metaphysical are not so exclusively human as they seem: human babies are not obviously self-aware, and Darwin thought that dogs, for example, could reflect on the past, altering their behaviour as they remembered earlier achievements. His strategy, rather than emphasizing the animal nature of humans, was to anthropomorphize animals. He ended *Descent* with praise for 'that heroic little monkey' who saved the life of his keeper.

But even if we are physiologically one with the rest of the animal world, and if the equipment we have to express emotion is not unique to us, could the emotions themselves – in particular those associated with morality, such as guilt and shame – be different? Blushing as a result of these complex emotions was thought to be uniquely human. Darwin could demonstrate that the circulation of the blood worked the same way in humans and animals, but that did not help: if humans, and only humans, were expressing something by a unique physiological response, then it could be argued that that something – a moral compass – had come from somewhere other than our shared common ancestry with animals. Even Alfred Russel Wallace was arguing by the late 1860s that the mental and moral qualities of humans lay outside the scope of natural selection and could be explained only by the intervention of a spiritual agency.

The fact that blushing is caused not just by shame but also by shyness was key for Darwin. For one thing, that made it very hard to explain blushing as a divinely designed mechanism for advertising guilt – why would a benevolent god make the shy suffer along with the wicked? It also suggested a possible origin for blushing, and a role for natural selection in its development. Darwin argued that we are acutely and instinctively aware of what others think of us – primarily their opinion of how we look (which, after all, is essential in the whole getting-a-mate thing), but also their opinion of our character and conduct – and we are particularly alert to and disturbed by the bad opinion of others. And if we believe others dislike something about our appearance or our conduct, we become self-conscious.

We can affect any part of our body, even if it is not normally under our voluntary control, by thinking about it (what Darwin called 'self-attention'), so, for example, thinking about eating may stimulate saliva or we may yawn when we see someone else yawning. In our interactions with one another, we naturally pay most attention to faces (Darwin thought this could have been reinforced in civilized society by generations of covering up the rest of our bodies), so our self-consciousness acts particularly on the circulation in our faces, relaxing the capillaries and resulting in a blush 'whenever we know, or imagine, that anyone is blaming, though in silence, our actions, thoughts, or character'; it is not guilt but 'the thought that others think us guilty' that makes us blush.[2]

Morality

'But as man gradually advanced in intellectual power ... his sympathies became more tender and widely diffused, so as to extend to the men of all races, to the imbecile, the maimed, and other useless members of society, and finally to the lower animals, – so would the standard of his morality rise higher and higher.'

Charles Darwin, *The Descent of Man*, 1871

A code of ethics seemed to many to be the ultimate defining characteristic of humanity. And it was here that Darwin was at his most uncompromising. He argued that human morality is a by-product of our social organization, and that a different social structure would lead to a different definition of morality. Morality is not absolute.

Humans are instinctively social animals, and, like other primates, within our immediate social circle we 'take pleasure in one another's company, warn one another of danger, defend and aid one another in many ways'.

As civilization progresses, that circle of sympathy will naturally extend beyond the immediate family, to embrace the tribe, the nation, the species or even beyond. The 'Golden Rule' (to 'do unto others as you would be done by') is, Darwin argued, a natural consequence of primate social order. But things would be very different for, say, bees, where the social instinct is to protect the queen, and where, therefore, killing any rivals would be morally justified.

Was Darwin a social Darwinist?

No, he was not. Darwin's privately expressed concerns about the consequences for the physical health of the human species of social movements that protected the weak – including trades unions and the co-operative societies – seem at first to be at odds with his public and very active involvement in local and national charities. But for Darwin the moral obligation to show sympathy towards the less fortunate, based as it was on fundamental social instinct, trumped fears about what the resultant (and, to his view, entirely necessary) humane policies might mean for the future. On Darwin's model of association, the further out one's sympathy extends beyond the immediate circle of family, to tribe, to nation and, ultimately, beyond one's own species, the higher is the moral achievement.

Race

Although by Darwin's own principle of divergence it seemed to him that the 'savage' races of humans, together with the most human-like apes, would inevitably die out, and those at the extremes of the spectrum (the 'civilized' humans and the lower apes) would survive, he was firmly convinced of the unity of the human species.

Both the Darwins and the Wedgwoods had a strong tradition of opposition to slavery. Darwin's grandfather Josiah Wedgwood I had been a friend and supporter of Thomas Clarkson, the founder of the Committee for the Abolition of the Slave Trade. A generation later, Darwin was sickened by the brutality he witnessed on the slave-owning plantations in South America; he was there in 1833 when slavery was outlawed throughout much of the British Empire, although it would not be completely abolished for another ten years. During the American Civil War (1861–5), Darwin sympathized strongly with the anti-slavery Unionist north.

One of Darwin's aims in *Descent* was to counter the argument that human races were separate species, with its obvious consequences for how those of different race were treated. Alongside natural selection, Darwin postulated a second mechanism for evolution – that of sexual selection. He argued that racial difference is the result of sexual selection, with mate choice coupled with varying aesthetic taste responsible for the divergence in physical appearance of different human groups.

Sexual selection

Female mate choice was crucial to Darwin's arguments about the development of otherwise useless physical adornment – bright colours, patterns that did not camouflage, ruffs, frills, overlarge antlers, heavy (but beautiful) peacock tails – and, in humans, of racial characteristics.

Alfred Russel Wallace had argued that the female birds of many species had become dull in appearance through natural selection as their role in protecting the young meant that they survived better when camouflaged. But this explanation would not do for Darwin because it could not account for the existence of beautiful plumage in the first place. And he needed to explain natural beauty as something other than a divine gift designed for human pleasure. He began by showing that, while individuals, or individual groups, may have strong opinions on what is beautiful, there is no universally accepted standard of beauty. And animals also exhibit preferences for particular colours, sounds and patterns, so aesthetic taste is not confined to humans. What we do appear to share is the instinct to enjoy novelty.

In species that have sex, Darwin realized that you either fight for it or flirt for it, and he was increasingly convinced that the more important of the two was flirting. He first published this idea in 1871 in *Descent*, the full title of which is *The Descent of Man and Selection in Relation to Sex*. Many secondary sexual characteristics, such as the large pincers in some male insects and crabs, allow the males either to grab and keep hold of females in order to mate with them, or to fight off other males. In the case of some striking characteristics, however, Darwin realized that sexual display itself had become the point. Although the combat role of antlers in the males of some species of deer was obvious, in others the antlers were so elaborate that they actually hampered the deer's ability to fight. They had instead become a pick-up line. Female deer were choosing to mate with the males they found the most attractive – in this case the ones who displayed the most elaborate antlers – and this process of female mate choice was influencing the characteristics that were passed on to the next generation.

This was unless you were a Victorian Western woman, of course: modern humans were the one species where Darwin thought female mate choice had ceased to operate.

Women

'... the happiness of our homes, would in this case greatly suffer.'

Darwin on women as breadwinners, in a letter to Caroline Kennard, 1882[3]

Darwin idolized his wife, appreciated his daughter's criticism of his work, recruited his nieces and daughter-in-law as unofficial research assistants, encouraged the scientific work of several women who wrote to him, and thought women morally superior to men. And yet he concluded in *Descent* that, although originally equal, men had become superior to women in imagination and powers of reasoning, and he believed to the end of his life that women were 'intellectually inferior'.[4]

He argued that the differing roles in sexual selection of males and females would tend, through natural selection, to increase not just the relative physical strength of males but also their intelligence. Essentially, he thought that stamina was a crucial part of genius and so males, who needed to be strong to survive, would also tend to be intelligent. He also thought that it made men naturally more aggressive, leaving women more tender and superior in virtue, but even in modern humans he thought the struggle to provide for a family would favour male intelligence. This was not an absolute difference, but the inevitable outcome of the way that human society had evolved, and it could, he thought, be reversed, but only by what to him seemed drastic and undesirable social change – the rigorous physical training and academic education of teenage girls, and the inclusion of women in the workforce. Even then, in order for the increased intelligence of these women to spread to the whole female population, Darwin pointed out that they would have to marry and have children – his unspoken implication being that this was unlikely.

Darwin believed that female mate choice had virtually ceased to operate in human societies once they became civilized, and had even been reversed, with men able to choose the most attractive women. He was particularly struck that in many societies it was the women who dressed up to attract a mate – often decorating their outfits with the same bright feathers that male birds used to attract females.

Do blondes have more fun?

Darwin encouraged inconclusive research into the increasing prevalence of dark hair in the British population, suspecting that it was driven by a male preference for dark-haired women, meaning that these women were more likely than their lighter-haired competitors to have sex – and children. So, according to Darwin, and assuming sex is fun, no, blondes don't have more of it.

7

Darwin and religion

'... no shadow of reason can be assigned for the belief that variations ... which have been the groundwork through natural selection of the formation of the most perfectly adapted animals in the world, man included, were intentionally and specially guided.'

Charles Darwin, *The Variation of Animals and Plants under Domestication*, 1868

Darwin said little in public about religion and he did not directly address the implications of his theories for religious belief in print until 1871, when he published *Descent*. Even then he did not say a great deal, but he had to say something because an apparently unique capacity for religious belief was one of the arguments used to claim 'special' creation for humans. *Descent* and *Expression* were designed to demonstrate that, far from being separately created, we are part of a continuum with all living things.

In writing about religion, Darwin was careful to make two points that helped defuse opposition to his views. The first was that the absence of a universal belief in God is wholly distinct from the 'higher' question as to whether a 'Creator or Ruler of the universe' actually exists. It may be possible to explain religious feelings without the necessary existence of a creator, but that does not mean one cannot exist, and the existence of a creator was, he said without commenting on his own position, accepted by 'the highest intellects that have ever lived' (though he later modified this to 'some' of the highest intellects).[1] Secondly, he asserted that belief in an omnipotent (even if possibly imaginary) god was a good thing: it was, he said, 'ennobling'. He was keen to promote the idea that belief in natural selection and in God were compatible, although 'different persons have different definitions of what they mean by God',[2] and he was for the most part cagey about his own definition.

Natural theology and the argument from design

The study of nature and the study of religion were generally thought to be in harmony. According to the tenets of what was known as 'natural theology', the study of one illuminated the study of the other, and the wonder and perfection of the natural world was evidence of divine design.

The most influential of these arguments, then and now, was most famously articulated by William Paley in his *Natural Theology: or, Evidences of the Existence and Attributes of the Deity, Collected from the Appearances of Nature*, published in 1802 – a book Darwin certainly read. Paley argued that anyone finding a watch lying on the ground and examining how intricately it was put together, and how much its operation depended on each part being just so, would assume the existence of an intelligent watchmaker; as living things are vastly more complex than watches, surely it stands to reason that they also must have an intelligent maker. To illustrate his argument in more detail, Paley chose to compare the eye with a telescope, the former seeming so much more perfect for its purpose.

It was on this battleground that Darwin chose most directly to counter-attack. He did not name Paley, but contemporaries would have had no difficulty in decoding the reference. Such a comparison was almost unavoidable, Darwin wrote, but why should we therefore conclude that even an organ as complex as the eye could only have been created by something analogous to human intelligence? Given sufficient time, and an initial thick layer of transparent tissue over a light-sensitive nerve, natural selection could produce all the various forms of eye in existence. Only if an organ could be identified that no number of intermediate steps could possibly create, could Darwin's theory be discounted. 'The eye to this day gives me a cold shudder,' he wrote to Asa Gray in 1860, 'but when I think of the fine known gradations, my reason tells me I ought to conquer the cold shudder.'[3]

To Darwin, much of the 'design' also just seemed plain bad. The wastefulness and cruelty of the natural world was to him an obvious and insurmountable objection to the whole idea of a Creator, or at least to a well-meaning one. Although agreeing explicitly with Paley that no feature of a living organism could come into existence (or at least not for long) that was other than beneficial to the creature itself, many such features were the opposite of beneficial to other animals: how could a beneficent God have designed wasps to lay their eggs in the bodies of other creatures that were then eaten alive by the hatchlings? How could cats have the instinct to play with mice? And what about when good features go bad, and changed circumstances lead – as they clearly had – to extinction?

Asa Gray (1810–88)

Darwin's most sustained discussion of religion for which we have evidence was with the Harvard Professor of Botany and devout Presbyterian Asa Gray. Around 300 surviving letters exist between Gray and Darwin, dating from 1854 to Darwin's death in 1881. One was the vital letter showing that Darwin had outlined the complete theory of natural selection by 1857 before receiving Alfred Wallace's paper. Gray promoted Darwin's ideas in North America, helping with the publication of a US edition of *Origin* and writing a series of influential reviews. Darwin seized on Gray's arguments as a defence against charges of irreligion and paid to have the reviews republished as pamphlets in Britain. Gray and his wife Jane visited the Darwins in 1868 and the two couples became friends.

When is a Creator not a Creator?

Darwin made a number of references in *Origin* to 'the Creator' in the context of the origin of life – which he never claimed to explain by natural selection – and of basic biological laws. In private he maintained that by 'creation' he really only meant '"appeared" by some wholly unknown process',[4] and this is supported by his daughter Henrietta's conviction that Darwin's occasional use of religious language was meaningless.

Darwin, essentially, was sugar-coating the pill. This kind of language helped to insulate both him and his ideas from charges of atheism. He had, he said, 'truckled to public opinion',[5] something he regretted in private but never wholly withdrew in public. Darwin also made the point that explanations of natural phenomena that invoked a Creator were essentially no explanation at all, and in the sixth and final edition of *Origin* he spelled out not once but twice that such arguments were unscientific.

The most striking of these references is an addition to the most famous passage in the whole book, the concluding paragraph in which Darwin condensed the whole of his argument into one highly descriptive and emotional passage. In the first edition this reads as follows:

'It is interesting to contemplate an entangled bank, clothed with many plants of many kinds, with birds singing on the bushes, with various insects flitting about, and with worms crawling through the damp earth, and to reflect that these elaborately constructed forms, so different from each other, and

dependent on each other in so complex a manner, have all been produced by laws acting around us ... There is grandeur in this view of life, with its several powers, having been originally breathed into a few forms or into one; and that, whilst this planet has gone cycling on according to the fixed law of gravity, from so simple a beginning endless forms most beautiful and most wonderful have been, and are being, evolved.'

Charles Darwin, *On the Origin of Species*, 1859, pp. 489–90

From the second edition onwards, the phrase 'originally breathed into a few forms or into one' was altered to read 'originally breathed by the Creator into a few forms or into one'. The language is evocative of the story of creation in the book of Genesis where God breathed life into Adam. To one sneering comment on his use of 'Pentateuchal' terms, Darwin responded that, although perhaps not appropriate in a purely scientific work, it helped convey that our ignorance is 'as profound on the origin of life as on the origin of force or matter'.[6]

A similar passage was also altered in the second edition of *Origin* from 'all the organic beings which have ever lived on this earth have descended from some one primordial form, into which life was first breathed' to 'first breathed by the Creator'. In this case the whole section was substantially rewritten in the third edition to make a much stronger case for a single primordial form as the ancestor of all living things, plant or animal, and the entire reference to the origin of life, not just to the Creator, quietly removed.

It is no accident that another mention of the Creator comes immediately after Darwin's only reference in *Origin* to the implications of his ideas for human evolution. After his famous statement that 'Light will be thrown on the origin of man and his history' and referring to the commonly held view that individual species had been separately created, he wrote:

'To my mind it accords better with what we know of the laws impressed on matter by the Creator, that the production and extinction of the past and present inhabitants of the world should have been due to secondary causes ... When I view all beings

not as special creations, but as the lineal descendants of some few beings which lived long before the first bed of the Silurian system was deposited, they seem to me to become ennobled.'

Charles Darwin, *On the Origin of Species*, 1859, pp. 488–9

This is essentially running interference: yes, natural selection applies to humans too, but don't worry, we can still see ourselves, if we wish, as the noble product of a divine plan.

The [god] is in the detail

By the time Darwin published *Variation*, his next major work after *Origin*, he could no longer completely evade the question of his position on divine intervention in the development, as opposed to the origin, of life – or, at least, not without appearing 'shabby'.[7]

But Darwin has a habit of revealing his hand only in the last paragraphs, or even the last sentence, of his books. You have to make it through 843 pages of detailed evidence, much of it printed in small type in order to squeeze in a mass of data, to find a statement that is as much conundrum as manifesto. It was written as an overt response to his friend Asa Gray, who reconciled a belief in God and in natural selection by suggesting that the variations on which natural selection operates had been 'led along beneficial lines', each specially ordained to produce, in particular, humans. Darwin pointed out that variation was not necessarily beneficial at all. Would a Creator preordain the particular variations that allow pigeon breeders, for example, to produce the absurdly misshapen pouters and fantails – variations that are individually not only useless but damaging? And if not, why should we accept that the particular variations that have led to humans were preordained? He concludes both the argument and the book with: 'Thus we are brought face to face with a difficulty as insoluble as is that of free will and predestination.' Many of those who asked Darwin for his personal views on the possible role of a Creator were referred to this passage as the best explanation they were going to get.

Of dogs and men: what Darwin said in *Descent*

Faced with the claim that humans are unique in having religious beliefs, Darwin, as usual, came up with more than one counter-argument.

Firstly, he pointed out that belief in a god, or even gods, is not universal, drawing on published anthropological evidence, but also on his own direct observation of the Fuegians he had encountered on the *Beagle* voyage who did not believe in anything 'we should call a god' or in the devil, and who had no religious rites.

Secondly, Darwin drew a careful distinction between belief in God and superstitious beliefs. A superstitious response postulating unseen spirits, to explain the otherwise apparently inexplicable, he did think was a 'near universal' tendency, but a tendency shared with animals. He described his own dog seeing a parasol blown around the garden by the wind and reacting as it would have done had there been a human intruder. The dog, according to Darwin, had unconsciously reasoned that the movement was caused by an invisible 'living agent'. And religious devotion in its compound of 'love, complete submission to an exalted and mysterious superior, ... strong sense of dependence, fear, reverence, gratitude' and 'hope for the future' had its parallel in the love of a dog for its master.

Hand in hand with a tendency to increased civilization and enhanced mental capacity, Darwin traced a natural progression from superstition through fetishism, then polytheism, and finally to monotheism. And as a bonus, the barbarous practices of primitive beliefs would, on this view, be abandoned as a result of increasing scientific enlightenment. This was a neat, almost subliminal, reconciliation of religion and science.

Darwin's personal beliefs

'What my own views may be is a question of no consequence to any one except myself.'

Letter to John Fordyce, 7 May 1879[8]

Darwin left little evidence for his beliefs, and he was generally reluctant to discuss them outside his family circle. His upbringing exposed him to liberal dissenting Christianity in the form of Unitarianism, and to atheism, but both of these were combined with a pragmatic acceptance of the established Anglican Church as an essential element of a stable society. Darwin's father and brother were atheists, while his mother and sisters remained practising Christians, but none of the family appears ever to have been fanatical in their views.

Darwin himself seems never to have wholeheartedly subscribed to the doctrines of the Anglican Church, and any belief in a personal god had gone by the time of his marriage in 1839. Although his views were not static, there is no evidence of a sudden crisis of faith, not even at the time of his daughter Annie's death in 1851, devastating though that was. A story that he had declared himself a Christian on his deathbed and renounced evolution has been well and truly debunked,[9] but the fact that such a myth could come into existence at all shows the intensity of interest in his personal beliefs during his lifetime and after, and the degree to which they are identified by some with the validity of his scientific theories.

In 1876 Darwin started *Recollections*, a private account, intended only for his family, of various aspects of his life, including his reflections on 'Religious belief'. In this he characterized himself first as a theist, then as an agnostic, but denied that he had ever been an atheist. He did not define what he meant by a 'theist' but said he felt occasionally compelled to see the wonder of the natural world as the result of a 'First Cause having an intelligent mind'. Even this impulse gradually, 'with many fluctuations', became weaker, and he concluded that he 'must be content to remain an Agnostic'. Darwin's friend Thomas Huxley had come up with the term 'agnostic' in order to draw a distinction between atheists (or 'materialists'), who positively deny the existence of any deity, and those who hold that anything beyond the material is simply unknowable – what their mutual friend Joseph Hooker described in 1871 as the distinction between 'un-belief' and 'a-belief'.[10]

After Darwin's death his family debated whether his privately expressed views in his written *Recollections* should be made public. Emma was concerned that, without the benefit of having heard him talk, anyone reading them would not appreciate the subtleties of his thought. Had he been writing for publication, she argued, Darwin would have been more careful in how he expressed himself and in particular would have avoided language that might offend the religious sensibilities of others. One of those strongly worded views was on the doctrine of eternal damnation for unbelievers, which to Darwin seemed so cruel that he could not understand how anyone could possibly wish Christianity to be true. Here Emma had the last word, writing in the margin after his death that very few people would call that doctrine Christian.

8

Darwinism after Darwin

'If I lived 20 more years, & was able to work, how I should have to modify the "Origin", & how much the views on all points will have to be modified. Well it is a beginning, and that is something.'

Letter to Joseph Hooker, 22 January 1869

1909: Darwin's centenary

Just as universities, and even governments, were sending envoys, messages and elaborate commemorative scrolls to Cambridge from all over the world in celebration of Darwin's hundredth birthday, a new word – genes – was coined that for a while seemed to undermine his evolutionary ideas.

The big gap in Darwin's own account of evolution (despite his much-loved Pangenesis) remained a convincing mechanism for how inheritance worked. At the same time as Darwin was conducting his cross-breeding experiments at Down, Gregor Mendel was conducting similar experiments at Brno (now in the Czech Republic). Mendel was an Augustinian friar, and a physics and philosophy teacher who had also studied agriculture. Between 1856 and 1863 he conducted a series of experiments on hybridization in peas that produced particularly interesting results, publishing his conclusions – which Darwin never saw – in 1866.

Mendel looked at characteristics where there were two possible alternatives – purple or white flowers, smooth or wrinkled seeds, yellow or green pods, and so on. He found that, although all the offspring of the first cross-bred generation exhibited one trait only (all had yellow pods, for instance), by self-pollinating these plants he got offspring where the lost trait reappeared in a consistent ratio of 1:3. He concluded that there were units of inheritance for each trait, which he called 'factors', each with two forms. Children inherited one form of each factor from each parent. The form of one trait – pod colour, for example – was independent of the form of the other traits – flower colour or the surface appearance of seeds. For each trait, however, one form was dominant and the other recessive. The recessive trait would appear only when inherited from both parents.

Mendel's work did not become generally known until 1900, when several scientists working independently rediscovered it, and it overturned the previously accepted view that inheritance acted by blending characteristics. In 1909 Mendel's 'factors' were renamed 'genes' ('forms' later became known as 'alleles').

1959: *Origin*'s centenary

Fifty years is a long time in evolutionary biology. Between 1909 and 1959 genetics had posed a number of challenges to natural selection. For a while it seemed to have become at best a secondary mechanism, with genetic mutation as the primary driver of species change, and Darwinian gradualism was replaced by saltation – the idea that evolution occurs not in tiny increments but in sudden leaps. By the 1940s, however, the 'modern synthesis' (a term coined by Thomas Huxley's grandson Julian) had forged an accommodation between genetics and natural selection, showing how the two operated together to drive evolution.

During Darwin's lifetime, the engineer Fleeming Jenkin had already raised a statistical problem with natural selection – the apparent probability that individual advantageous variations will simply be wiped out by chance accidents before they can affect the population as a whole: on the model of blending inheritance, which was the one generally accepted in Darwin's time, they would simply be swamped.

In the 1920s and 1930s the population geneticists Ronald Fisher, Sewall Wright and J. B. S. Haldane used a statistical approach to combine Mendelian inheritance with natural selection. They showed that, although genetic mutations are generally recessive and disadvantageous, where they are advantageous, natural selection allows them to spread. One irony was Fisher's demonstration that Malthusian population growth is not necessary after all for natural selection to operate: only heritable variation, differential fitness, and correlation between these two.

The Russian-born geneticist Theodosius Dobzhansky, working in the US during the 1930s, built on their research to demonstrate the role of genes in the creation of new species. By studying geographically isolated populations of wild fruit flies, he showed that small genetic mutations were more common than anyone had realized and could spread rapidly through the local community. Most of the mutations had little effect on the external characteristics of the flies, but where the newly mutated genes dominating one localized population were incompatible with those in other localized populations, distinct species emerged that were incapable of interbreeding. In the following decade, the ornithologist Ernst Mayr put forward the theory that recognizably different sub-species of birds inhabiting different areas within the species' geographical range were also evidence of localized genetic variation. In this case, however, because the local populations with their differing characteristics (longer tails, bigger crests or different beaks, for example) were not completely isolated from one another, neighbouring groups continued to interbreed. Mutually incompatible mutations were not able to spread, and the sub-species did not evolve into separate species.

2009: Darwin's bicentenary

By 2009 the application of Darwin's ideas had been explored so widely that the international conferences held all over the world were as much about philosophy and politics as about biology. Darwin's name appeared in the popular press more than in academic journals, and he even made it to the big screen with the biopic *Creation*. Most modern challenges to Darwinian natural selection come not from within science but from religious fundamentalists.

Between 1959 and 2009 the operation of natural selection in a natural population was demonstrated for the first time by H. B. D. Kettlewell and confirmed by Michael Majerus. They studied the peppered moth and showed that the relative numbers of its lighter and darker form changed markedly in response to changes in the environment. The darker form was better camouflaged against predators during times of industrial pollution, but declined again once pollution levels started to fall: that form may well be extinct before the bicentenary of *Origin* in 2059.

100 IDEAS

This 100 ideas section gives ways you can explore the subject in more depth. It's much more than just the usual reading list.

Five Darwin food facts

1 **India Pale Ale** – Darwin drank this at lunch following medical advice from his father, and it suited him 'very well'. He wasn't so happy about following the advice to avoid sugar (*Correspondence* vol. 2, letter to Susan Darwin [27 November 1844?]).

2 **Rice** – the only recipe by Charles in his wife's handwritten cookery book is one for boiled rice. The book is in the Darwin Archive of Cambridge University Library, and there is a published version adapted for modern cooks (D. Bateson and W. Janeway, *Mrs Charles Darwin's Recipe Book Revived and Illustrated* (New York: Glitterati, 2008)). Other recipes include buttered eggs, French ragout of mutton, stewed spinach, and Nesselrode pudding (a brandy ice cream with dried fruit).

3 **Coffee** – Darwin drank coffee every day at university but gave it up for 20 years and only started drinking it again in his fifties when it seemed to help with headaches and wind.

4 **Potatoes** – In the 1870s, both for humanitarian reasons and because he was interested in the light it might throw on hybridity, Darwin helped support attempts by the Irishman James Torbitt to develop blight-resistant potatoes.

5 **Owl** – Darwin and his student friends started the 'Glutton Club' to experiment with 'birds & beasts which were before unknown to human palate' (see *Correspondence* vol. 1, letter from Frederick Watkins, [18 September 1831], n. 1). It supposedly ended when they tried eating an old brown owl. It didn't agree with them.

Ten ideas for places to visit

6 **Shrewsbury** – where Darwin grew up. The Mount, the house where he was born, is now a tax office.

7 **Cambridge** – Specimens brought back from the *Beagle* voyage are on show at the Sedgwick Museum of Earth Sciences (which has a permanent Darwin exhibit) and the University Museum of Zoology, which also has Darwin's beetle collection (see www.museum.zoo.cam.ac.uk/site_resources/darwin_guide.pdf – closed until 2016). Darwin's rooms in Christ's College are still occupied and are not often open to the public, but there is a statue of Darwin as a young man, complete with a couple of beetles, in the college grounds. Darwin's lodgings in Sidney Street (now over a branch of Boots the Chemist) and in Fitzwilliam Street are both marked with plaques. Houses occupied by Francis Darwin and Emma Darwin are part of Darwin

College and Murray Edwards College, and several Darwins are buried in Histon Road Cemetery.

8 **London** – Darwin was a member of the Linnean Society, the Geological Society, the Royal Society and the Athenaeum; plaques mark where he lived in Upper Gower Street and Great Marlborough Street. Find the statue of him in the Natural History Museum, and visit his grave in Westminster Abbey.

9 **The Darwin Walk** – near Wentworth Falls in the Blue Mountains, New South Wales, Australia.

10 **The Galápagos** – no list would be complete ... but it will cost you.

11 **Glen Roy, Scotland** – For a downloadable guide by Cambridge geologist and historian Martin Rudwick, see www.darwinproject.ac.uk/darwin-glen-roy.

12 **Down House** – the Darwins' house in Downe, Kent, now run by English Heritage. The downstairs rooms, including Darwin's study, have been restored and there are exhibitions and a learning centre upstairs. The hothouses are still there, and the gardens, including the kitchen garden, have been replanted as they were in the 1870s. Walk around Darwin's thinking path, and see a recreation of his worm-stone experiment (www.english-heritage.org.uk/visit/places/home-of-charles-darwin-down-house/)

13 **Ilkley, Yorkshire** – where Darwin was staying when *Origin* was published. See www.bshs.org.uk/travel-guide/ilkley

14 **Bahia, Brazil** – where Darwin first set foot in South America in February 1832.

15 And one to avoid maybe – the island of **Chiloé** off the coast of Chile, which Darwin called a 'miserable hole', complaining that it never stopped raining there.

Five Darwin jokes (by and about)

16 Martin Rowson cartoon in the *Guardian*, 17 December 2011, on the death of the 'unshakable secularist', Christopher Hitchens: a white-bearded figure at the gates of Heaven welcomes an aghast Hitchens with the words 'Relax! I'm Charles Darwin!' (www.theguardian.com/commentisfree/cartoon/2011/dec/16/martin-rowson-christopher-hitchens-cartoon)

17 One day the zookeeper noticed that the orang-utan was reading two books – the Bible and Darwin's *On the Origin of Species*. In surprise he asked the ape, 'Why are you reading both?' 'Well,' said the orang-utan, 'I want to know if I am my brother's keeper or my keeper's brother.'

18 Joseph Hooker to Charles Darwin: 'Blood, Blunt, Brains, Beauty – these are all good things, of use to the organism possessing them, & hence sought after by all human organisms, & their accumulation, by Natural Selection, must culminate in an Aristocracy – or there is no truth in Darwinism.' (*Correspondence* vol. 10, letter from J. D. Hooker [23 March 1862]).

19 If Darwin was right, you will probably figure it out in a few million years.

20 Thomas Henry Huxley, as reported by Darwin: 'Genesis is difficult to believe, but Pangenesis is a deuced deal more difficult.' (*Correspondence* vol. 16, letter to J. V. Carus, 21 March [1868]).

Ten things that arrived in the post

21 **Locust dung**

22 **Bees (dead)**

23 **Rabbits (dead and alive)**

24 **Watercolours**

25 **Worm casts**

26 **Beard hair**

27 **Bird skins**

28 **Photographs** – Charles Lutwidge Dodgson (aka Lewis Carroll) sent Darwin a photograph of a young girl, Flora Rankin, after reading Darwin's book *Expression*.

29 **Butterfly wings**

30 **Postage stamps** – begged for Darwin's son Leonard, who collected them

Ten people Darwin corresponded with

31 **Lydia Becker**, a leader of the women's suffrage movement as well as a botanist and astronomer. She founded the Ladies' Literary Society in Manchester, the real purpose of which was discussion of science; Darwin sent her one of his botanical research papers to read at a meeting.

32 **Mary Boole**, wife of the mathematician George Boole (of Boolean logic) and a mathematician in her own right. She asked Darwin for his views on the compatibility of science and religion.

33 **James Crichton-Browne**, superintendent of Britain's largest mental asylum in Wakefield, Yorkshire, with over a thousand patients. He sent Darwin photographs of patients and case notes, and advised him on the section in *Expression* on blushing. His letters to Darwin are almost the only direct evidence of this important stage in Crichton-Browne's own career.

34 **George E. Harris**, a 'poor tailor' with a wife and four children to support, wrote to Darwin asking for a copy of *Origin* as he couldn't afford to buy one himself. We don't know whether Darwin sent him one.

35 **John Lubbock**, who was eight when the Darwins moved in next door. He became a favourite of Darwin's, and was among his closest scientific friends. An anthropologist and an entomologist as well as a banker and Member of Parliament, he kept a pet wasp, saved the Avebury stone circle for the nation, and established bank holidays.

36 **St George Jackson Mivart**, a Roman Catholic zoologist, who first offended Darwin by appearing to be friendly while publishing anonymous and highly critical reviews of his work. When he wrote a review of a paper by one of Darwin's sons implying that it condoned immoral behaviour, Darwin angrily ended their correspondence.

37 **Fritz Müller**, described by Darwin as the 'prince of observers', who emigrated to Brazil while his brother Hermann stayed in Germany. They had a three-way correspondence with Darwin, both sending observations of plants and animals, and trying various experiments. The brothers also made important scientific discoveries in their own right based on their acceptance of natural selection.

38 **Fanny Owen**, whose letters Darwin carried when he left England on HMS *Beagle.* The letters from Fanny to her 'dear Postillion' urged him to come home quickly and were signed his 'ever faithful Housemaid'. The first letters from home that he got in South America told him that Fanny was engaged to marry someone else; he was heart-broken.

39 **William Bernhard Tegetmeier**, a journalist and pigeon fancier. Tegetmeier and Darwin exchanged hundreds of letters over more than 25 years, with Tegetmeier providing

Darwin with information and specimens, and conducting experiments on his behalf.

40 **William Darwin Fox**, a cousin who became close friends with Darwin when they were students at Cambridge. Fox and Darwin wrote to one another sporadically for the rest of Darwin's life. Their 224 surviving letters are among the most affectionate and revealing of any of Darwin's correspondence. A good example is Darwin's letter to Fox of 7 March 1852. You can read it on the Darwin Correspondence Project website (www.darwinproject.ac.uk) and in vol. 5 of the *Correspondence*.

Ten things for and about kids

41 One of the best illustrated books for younger children: **P. Sìs, *The Tree of Life*** (Farrar, Straus & Giroux, 2003). See www.petersis.com/content/tree.html).

42 In the Discworld Science series for teens and older: **T. Pratchett, I. Stewart and J. Cohen, *Darwin's Watch*** (London: Ebury Press, 2005).

43 For Darwin experiments to do at home, see the **Darwin Correspondence Project schools' resources** (www.darwinproject.ac.uk/schoolsresources/science/).

44 Only around 40 of the original manuscript pages of *Origin* survive, and they were mostly kept for the sake of **drawings on the back by Darwin's children**, who were given the sheets as scrap. See the drawings (and the manuscript of *Origin*) at www.amnh.org/our-research/darwin-manuscripts-project/featured-collections/children-s-drawings-stories2.

45 **Read Charles and Emma Darwin's own observations of their children** on the Darwin Correspondence Project website (www.darwinproject.ac.uk/observations-on-children)

46 Find **games, videos, teaching resources and activities** by the Charles Darwin Trust at www.charlesdarwintrust.org/content/19/darwin-inspired-learning.

47 The Linnean Society has **teaching resources for A-level students** on variation and adaptation, selection, genetics, classification and phylogeny. See www.linnean.org/Education+Resources/Secondary_Resources/darwin_inspired_learning).

48 Not remotely accurate, but fun: watch ***Pirates! In an Adventure with Scientists!*** by Aardman Animation Studios (of Wallace and Gromit fame). Charles Darwin is voiced by David Tennant. It is a little-known fact that Gromit shares a birthday with Charles Darwin (12 February).

49 Take them to **Down House** (see under 'Places to visit').

50 You can explore the geological tools and techniques Darwin used in the online activities at the **Sedgwick Museum of Earth Sciences in Cambridge** (www.sedgwickmuseum.org/index.php?page=darwin).

Twenty ideas for finding out more

51 The quickest way to read Darwin's *Journal of Researches (Voyage of the Beagle), Origin, Descent,* and his autobiographies, is in the edition by **James Secord, *Evolutionary Writings*** (Oxford: Oxford World's Classics, 2008).

52 **Read Darwin's other publications,** including *Expression* and *Earthworms* online at http://darwin-online.org.uk.

53 See Darwin's manuscripts online and read introductions and transcriptions at the **Darwin Manuscripts Project**, American Museum of Natural History (www.amnh.org/our-research/darwin-manuscripts-project).

54 **A comprehensive two-volume biography:** J. Browne, *Charles Darwin: Voyaging* (New York: Alfred A. Knopf; London: Cape, 1995), and *Charles Darwin: The Power of Place* (London: Pimlico, 2002).

55 **J. Browne, *Charles Darwin's* Origin of Species: *A Biography*** (London: Atlantic Books, 2006).

56 See **Conrad Martens' sketchbooks** from the *Beagle* voyage and other Darwin manuscripts and letters at the Cambridge Digital Library (http://cudl.lib.cam.ac.uk/).

57 Family reminiscences by Darwin's granddaughter Gwen Raverat: ***Period Piece: A Cambridge Childhood*** (London: Slightly Foxed, 2013).

58 **M. Ruse (ed.), *The Cambridge Encyclopaedia of Darwin and Evolutionary Thought*** (Cambridge: Cambridge University Press, 2013).

59 **G. Beer, *Darwin's Plots: Evolutionary Narrative in Darwin, George Eliot and Nineteenth-century Fiction*** (Cambridge: Cambridge University Press, 2009).

60 **R. Keynes, *Annie's Box: Charles Darwin, His Daughter and Human Evolution*** (London: Fourth Estate, 2001).

61 **T. Lewens, *Darwin*** (Abingdon: Routledge, 2006).

62 **S. Herbert, *Charles Darwin, Geologist*** (Ithaca: Cornell, 2005).

63 **R. Stott, *Darwin and the Barnacle*** (London: Faber, 2003).

64 **Christ's College Cambridge** has useful online resources on Darwin: http://darwin200.christs.cam.ac.uk/pages/.

65 And so does the **Wellcome Trust**, including the video *The Tree of Life* narrated by Sir David Attenborough (www.wellcome.ac.uk/Funding/Public-engagement/Funded-projects/Major-initiatives/Darwin-200/index.htm).

66 Read **Charles Lyell's *Principles of Geology*** in the Penguin Classics edition edited by James Secord (2007).

67 See **images of plants collected by Darwin** both before and during the *Beagle* voyage: http://cambridgeherbarium.org/collections/darwin-specimens/darwins-plants-at-cambridge.

68 **Find out what Darwin read:** see www.darwinproject.ac.uk/what-darwin-read for his student book list, a list of books on the *Beagle*, and his reading notebooks.

69 Many of the books Darwin had in his own library are available online through the **Biodiversity Heritage Library site** (www.biodiversitylibrary.org/browse/collection/darwinlibrary).

70 Get the books Darwin had on the *Beagle* for yourself from Cambridge University Press in their '**Cambridge Library Collection**' series (www.cambridge.org).

Ten pop culture references

71 ***Creation*** the movie: loosely based on the book *Annie's Box* by Darwin's great-grandson Randal Keynes. Stars Paul Bettany as Darwin.

72 Other actors who have played Charles Darwin: David Tennant (the tenth Dr Who) voiced him in *Pirates! in an Adventure with Scientists!*; Terry Molloy (Mike Tucker in *The Archers*, and Davros in *Dr Who*) in the stage play *Re:Design* by Craig Baxter; Henry Ian Cusick (star of *Lost*) in the TV drama *Darwin's Darkest Hour*; Edi Gathegi (sort of) as Armando Muñoz aka Darwin in *X-Men: First Class*.

73 The lyrics to the REM track '**Man in the Moon**': '... Mr Charles Darwin had the gall to ask'.

74 **Darwin dolls and puppets:** see the Unemployed Philosophers Guild (www.philosophersguild.com).

75 **'Hey Charly'**: Suzy Quatro's 1991 song for Charles Darwin.

76 **The paintings of the contemporary American artist Alexis Rockman,** inspired by Ernst Haeckel and ideas of evolution.

77 In the movie ***Inherit the Wind*** a teacher is prosecuted for teaching Darwin's theories. Based on the 1925 'Scopes Monkey Trial', with Spencer Tracy and Gene Kelly, and directed by Stanley Kramer. Bruce Springsteen recorded a track **'Part Man, Part Monkey'** also about the Scopes trial.

78 Darwin will be replaced by Jane Austen as the face on the UK ten-pound note.

79 'Like Darwin's finches, we are all slowly evolving': voiceover, ***Best Exotic Marigold Hotel***.

80 ***The Big Bang Theory***, series 4 episode 18, 'The Prestidigitation Approximation': Leonard tries to explain to Penny that his new girlfriend wants him to end their friendship by referring to Darwin's finches. She doesn't get the point. It's not surprising.

Five things named after Darwin

81 **Darwin, Australia**

82 ***Calceolaria darwinii*** (now *Calceolaria uniflora*), or Darwin's slipper flower. Darwin discovered this plant in Tierra del Fuego and Conrad Martens made a watercolour sketch of it in one of his sketchbooks.

83 **Darwin Island**: among the smallest in the Galápagos archipelago with an area of just one square kilometre.

84 *Berberis darwinii*, Darwin's barberry. A common garden shrub in the UK.

85 *Beagle 2*, the Mars lander, was named after HMS *Beagle*. It was launched in June 2003 and was a British-led effort to land on Mars as part of the European Space Agency's Mars Express Mission. Their website quotes Darwin's father's objection to the original *Beagle* voyage: 'A wild scheme, it would be a useless undertaking.' http://beagle2.open.ac.uk/index.htm.

Ten literary works Darwin read

86 *Jane Eyre* by Charlotte Brontë

87 *Quentin Durwood* by Walter Scott

88 'Some of Shelley's poems'

89 *The Vicar of Wakefield* by Oliver Goldsmith

90 Samuel Pepys's *Diary* – 'skimmed one volume'

91 *Gulliver's Travels* by Jonathan Swift

92 *Robinson Crusoe* by Daniel Defoe

93 *Don Quixote* by Miguel de Cervantes

94 *Martin Chuzzlewit* by Charles Dickens

95 Byron's *Childe Harold*

Four things Darwin said (and one he didn't)

Darwin did say:

96 'I am quite conscious that my speculations run quite beyond the bounds of true science' (*Correspondence* vol. 6, letter to Asa Gray, 18 June 1857). This is often quoted out of context to argue that Darwin himself doubted his theories, but it isn't about the theory of natural selection as a whole, only one small problem in looking at the distribution of plant species. Darwin and Gray had been discussing how to account for species for which there are no, or few, other closely related species. Darwin speculated that these 'disjoined species' would be found to come from genera with very few species in total. But since he had not been able to test this against much evidence, it *was* just speculation and therefore 'unscientific'. This is Darwin being a good scientist, careful to understand the limits of the evidence available to him.

97 'You will surely find that the greatest pleasure in life is in being beloved; & this depends almost more on pleasant manners, than on being kind with grave & gruff manners' (*Correspondence* vol. 5, to his son William Erasmus Darwin, 3 October 1851).

98 'I fear parts are too like a Sermon: who wd ever have thought that I shd. turn parson?' To his daughter Henrietta, who was reading the manuscript of *Descent of Man* (*Correspondence* vol. 18, letter to Henrietta Darwin [8 February 1870]).

99 'To receive the approbation & sympathy of one's fellow-workers in the acquisition of knowledge is the highest possible reward which any man ought to desire.' To the masters of Greiz College in reply to their congratulations on his 70th birthday (American Philosophical Society (Getz 11884)).

And he definitely didn't say:

100 'It is not the strongest of the species that survives, nor the most intelligent that survives. It is the one that is most adaptable to change.' This has been traced to a paraphrase of a 1963 book by the management guru Leon C. Megginson and is one of the most quoted misquotes on the Web (see the blog by Nick Matzke at www.pandasthumb.org). It is also set in the floor of the California Academy of Sciences, though no longer attributed to Darwin. See more like this at www.darwinproject.ac.uk/six-things-darwin-never-said.

Abbreviations used for Darwin's writings

Correspondence
F. Burkhardt et al. (eds), *The Correspondence of Charles Darwin* (Cambridge: Cambridge University Press, 1985–).

Descent
The Descent of Man and Selection in Relation to Sex (London: John Murray, 1871).

Expression
The Expression of the Emotions in Man and Animals (London: John Murray, 1872).

Origin
On the Origin of Species by Means of Natural Selection, or the Preservation of Favoured Races in the Struggle for Life (London: John Murray, 1859).

Recollections
'Recollections of the Development of My Mind and Character, 1876–81', in J. A. Secord (ed.), *Evolutionary Writings* (Oxford: Oxford University Press, 2008).

Variation
The Variation of Animals and Plants under Domestication (London: John Murray, 1868).

Notes

Chapter 1

1 Lists are now in Cambridge University Library MS DAR 210.8: 1 and 2; see www.darwinproject.ac.uk/darwins-notes-on-marriage.

2 *Correspondence* vol. 2, letter to Susan Darwin [late July–August 1842].

3 *Correspondence* vol. 10, letter to John Lubbock, 5 September [1862].

4 Jane Gray to Susan Loring, 28 October–2 November 1868 (Harvard University, Archives of the Gray Herbarium, Asa Gray Personal Papers, AG-Bio, Box G, Folder 15); see also www.darwinproject.ac.uk/jane-gray-on-down.

5 Ibid.

6 *Correspondence* vol. 3, letter to Joseph Dalton Hooker [11 January 1844]; see the original at http://cudl.lib.cam.ac.uk/view/MS-DAR-00114-00003/1.

7 R. Chambers, *Vestiges of the Natural History of Creation* (London: John Churchill, 1844).

8 C. Darwin and A. R. Wallace, 'On the tendency of species to form varieties; and on the perpetuation of varieties and species by natural means of selection ... Communicated by Sir Charles Lyell ... and J. D. Hooker'

(read 1 July 1858), *Journal of the Proceedings of the Linnean Society* (Zoology) 3 (1859): 45–62.

9 *Correspondence* vol. 7, letter from Adam Sedgwick, 24 November 1859.

Chapter 2

1 For a history of the word 'evolution', see the *Oxford English Dictionary* entry, and P. J. Bowler, 'The Changing Meaning of "Evolution"', *Journal of the History of Ideas* 36 (1975): 95–114.

2 A. R. Wallace, 'On the Law Which Has Regulated the Introduction of New Species', *Annals and Magazine of Natural History*, 2nd series, 16 (1855): 184–96.

3 P. Matthew, *On Naval Timber and Arboriculture; with critical notes on authors who have recently treated the subject of planting* (London: Longman, Rees, Orme, Brown, & Green; Edinburgh: Adam Black, 1831).

4 Cambridge University Library, MS DAR 29.2: 74.

Chapter 3

1 *Correspondence* vol. 20, letter to Hermann Müller [before 5 May 1872].

2 For Francis Darwin's description of his father's study and work habits, see *The Life and Letters of Charles Darwin*, 1: 108–61 (London: John Murray, 1887–8).

3 *Correspondence* vol. 1, letter to J. S. Henslow, 24 July–7 November 1834.

4 Cambridge University Library, MS DAR 29.3: 78.

5 *Correspondence* vol. 1, letter from Thomas Campbell Eyton, 12 November 1833.

6 Letter to W. T. Thiselton-Dyer, 14 July 1878 (Royal Botanic Gardens, Kew: Thiselton-Dyer, W. T. letters: 135–6).

7 F. Burkhardt et al. (eds), *The Correspondence of Charles Darwin* (Cambridge University Press, 1985–).

8 Cambridge University Library, MS DAR112: B9-B23, Recollections of George Darwin (see darwin-online.org.uk/).

9 Jane Gray to Susan Loring, 28 October–2 November 1868 (Harvard University, Archives of the Gray Herbarium, Asa Gray Personal Papers, AG-Bio, Box G, Folder 15); see also www.darwinproject.ac.uk/jane-gray-on-down.

Chapter 4

1 *Correspondence* vol. 1, letter to Catherine Darwin, 22 May–14 July 1833.

2 *Correspondence* vol. 1, letter to Sarah Elizabeth (Elizabeth) Wedgwood, 28 August 1837.

3 *Correspondence* vol. 11, letter to John Scott, 6 June 1863. On Darwin's difficulties with terminology, see Beer, G., *Darwin's Plots: Evolutionary Narrative in*

Darwin, George Eliot and Nineteenth-century Fiction (Cambridge University Press, 2009).

Chapter 5

1 *Correspondence* vol. 9, letter to T. F. Jamieson, 6 September 1861.

2 B. Rosen, 'Darwin, Coral Reefs, and Global Geology', *BioScience* 32: 6 (1982): 519–25 (see darwin-online.org.uk).

3 Darwin, C., *On the Movements and Habits of Climbing Plants* (London: Longman, 1865), p. 118.

4 *Correspondence* vol. 11, letter to Asa Gray, 4 August 1863.

5 *Correspondence* vol. 22, letter to Asa Gray, 3 June 1874.

6 C. Darwin, *The Power of Movement in Plants* (London: John Murray, 1880).

7 *Correspondence* vol. 16, letter to H. W. Bates, 22 February 1868.

8 *Correspondence* vol. 13, letter from T. H. Huxley, 16 July 1865.

Chapter 6

1 Quoted in A. Desmond, *Huxley: The Devil's Disciple* (London: Michael Joseph, 1994), p. 241.

2 *Expression*, p. 346

3 Letter to C. A. Kennard, 9 January 1882, Cambridge University Library MS DAR 185: 29.

4 Ibid.

Chapter 7

1 *Descent*, 1st and 2nd edns.

2 From Emma Darwin writing on Charles's behalf to N. A. von Mengden, 8 April 1879 (Cambridge University Library MS DAR 139.12: 14).

3 *Correspondence* vol. 8, letter to Asa Gray, 8 or 9 February 1860. 'On the evolution of the eye', see *Origin*, pp. 186–7.

4 *Correspondence* vol. 11, letter to J. D. Hooker, 29 March 1863.

5 Ibid.

6 *Correspondence* vol. 1, letter to *The Athenaeum*, 18 April 1863.

7 *Correspondence* vol. 15, letter to J. D. Hooker, 8 February 1867.

8 Letter from Charles Darwin to John Fordyce, 7 May 1879, Linnean Society of London (Quentin Keynes Collection).

9 J. R. Moore, *The Darwin Legend* (London: Hodder & Stoughton, 1995).

10 *Correspondence* vol. 19, letter from J. D. Hooker, 5 August 1871.

Index

All That Matters books are written by the world's leading experts, to introduce the most exciting and relevant areas of an important topic to students and general readers.

From Bioethics to Muhammad and Philosophy to Sustainability, the *All That Matters* series covers the most controversial and engaging topics from science, philosophy, history, religion and other fields. The authors are world-class academics or top public intellectuals, on a mission to bring the most interesting and challenging areas of their subject to new readers.

Each book contains a unique '100 ideas' section, giving inspiration to readers whose interest has been piqued and who want to explore the subject further.